Calculus

The author advising the 1991 "SMALL" undergraduate research Geometry Group, which incidentally edited the work which is described on page 67. At the head of the table, the author; to his right, Hugh Howards and Steve Root; to his left, Kathryn Kollett, Thomas Colthurst and Holly Lowy (not visible Chris Cox and Joel Foisy). The Williams College "SMALL" Project is a National Science Foundation site for Research Experiences for Undergraduates. (Photo by Cheryl LeClaire.)

Calculus

Frank Morgan
Department of Mathematics and Statistics
Williams College
Williamstown, Massachusetts

Succeeding Calculus Lite.

ISBN-13: 978-1478356882
ISBN-10: 147835688X

Cover picture shows one of infinitely many optimal convex pentagonal planar tilings. Figures 1.1, 8.1, 8.2, 8.5, 8.9, 8.10, 22.1 and 30.1 by James F. Bredt.

Manuscript typed by Manuel Alfaro.

Student editorial committee: Cordelia Aidkin, Joe Corneli, Chris Cox, Stephen Fiedler, Joel Foisy, Amy Huston, Kathryn Kollett, Holly Lowy, Jonathan Mattingly, Stephen Root, Eric Swanson.

This book is dedicated to all calculus teachers, especially my brother David.
Here David and I are leaving for school.
(Photograph courtesy of the Morgan family; taken by the author's father, Mr. Frank E. Morgan.)

Contents

Preface

Many students and faculty spend a lot of time wading through fat calculus books. This lean text covers single-variable calculus in 300 pages by

1. getting right to the point, and stopping there,

2. introducing some standard preliminary topics, such as trigonometry and limits, by using them in the calculus.

Lots of attention goes to the most important topics, such as maxima-minima problems, graphing, and a natural proof of the fundamental theorem. Maximizing the areas of rectangles leads into a discussion of current research on soap bubbles, some by undergraduates. Integration by table, partial fractions, integration by parts, and numerical methods get the burst of attention they deserve. Focusing partial fractions on distinct, linear factors provides the theory and ninety percent of the applications without the time-consuming algebra.

Hard exercises are sometimes marked with an asterisk.

Alan Durfee[1] and colleagues at Mount Holyoke College supplement the text with technology and interesting projects, such as "How the compound eye of the bee is designed so that it has the best vision." Edward Burger at Williams[2], whose students study this text and his own video calculus course from Thinkwell.com for homework, has class time free from lecture for more interactive activities.

Students may also enjoy the news and contests in my web column at MathChat.org.

The result is this useful, short, inexpensive calculus book. It has no special perspective, no hidden agenda, no other purpose. The course can assume your agenda, rather than that of the book.

---

[1] adurfee@MtHolyoke.edu
[2] Edward.Burger@williams.edu

Introduction

The history of humanity is the *intellectual* drama of mankind's finding its place in the universe. In this historical drama calculus has played a central role.

Poring over the tomes of data that Tycho Brahe (1546–1601) collected over a lifetime of planetary observations, Johannes Kepler (1571–1630) noticed certain patterns. Figuring out how the orbits would look from above, Kepler deduced that the planets were moving in ellipses. Then Sir Isaac Newton (1642–1727, pictured on the next page) invented the calculus in order to relate Kepler's Laws to his own embryonic $F = ma$ and law of gravitation; the conclusions were published in his celebrated *Principia*.

Meanwhile on the continent, Baron Gottfried Wilhelm Leibniz (1646–1716) apparently also invented the calculus. Many years of argument over how to divide the credit followed.

The long story had a dramatic episode this past century. The scientific historian Emil Fellmann 4 had given a lifetime's attention to the controversy. One day a book dealer approached him with yet

I Vanderbank pinxit 1725   Geo. Vertue Sculpsit
1726 *Philosphiœ Naturlis Principia Mathematica* by Isaaco Newtono,
Editio tertia aucta & emendata, Londini: Apud Guil. & John Innys,
Regiæ societas typographos, MDCCXXVI.

Sir Isaac Newton (1642-1727) invented the calculus in order to relate Kepler's Laws
to his own embryonic $F = ma$ and law of gravitation, the conclusions published
in his celebrated *Principia*. This picture comes from a copy of the *Principia* at
the Institute for Advanced Study, Princeton. For wonderful hospitality during
my year there and especially for this picture, I would like to thank the Institute,
and Elliott Shore, librarian.

another copy of Newton's *Principia*, apologizing for its condition. As Fellmann paged through it, he had a realization of the sort few human beings ever know: he realized it was Leibniz's own copy, annotated in Leibniz's own hand! (See next page.)

It turned out that the copy had lain unnoticed in the University library at Göttingen since 1749. In 1926, considered "spoiled" by the annotations, it was sold, in favor of a clean copy. Today, thanks to Fellmann, it rests at the Bibliotheca Bodmeriana in Cologny, Switzerland.

Leibniz's annotations show his attempt to reformulate Newton's calculations in a different notation of his own with which he was already familiar. This evidence substantiates Leibniz's claim to an independent invention of the calculus.

Fellmann kindly provided the interesting page with Leibniz's annotations. Herman Karcher first told me the story, and Stefan Hildebrandt introduced me to Fellmann.

May this book help you understand something of the calculus and perhaps glimpse something of the drama.

Frank Morgan
Williamstown, Massachusetts
Frank.Morgan@williams.edu
www.williams.edu/Mathematics/fmorgan

Tangentes, & similia peragendi, quæ in terminis surdis æque ac
in rationalibus procederet, & literis transpositis hanc sententiam
involventibus [Data æquatione quotcunq; fluentes quantitates
involvente, fluxiones invenire, & vice versa] eandem celarem : re-
scripsit Vir Clarissimus se quoq; in ejusmodi methodum incidisse,
& methodum suam communicavit a mea vix abludentem præter-
quam in verborum & notarum formulis.   Utriusq; fundamentum
continetur in hoc Lemmate.

Prop. VIII. Theor. VI.

*Si corpus in Medio uniformi, Gravitate uniformiter agente, recta as-
cendat vel descendat, & spatium totum descriptum distinguatur
in partes æquales, inq; principiis singularum partium (addendo
resistentiam Medii ad vim gravitatis, quando corpus ascendit,
vel subducendo ipsam quando corpus descendit) colligantur vires
absolutæ; dico quod vires illæ absolutæ sunt in progressione Geo-
metrica.*

Exponatur enim vis gravitatis per datam lineam AC; resisten-
tia per lineam indefinitam AK; vis absoluta in descensu corporis
per differentiam KC; velocitas corporis per lineam AP (quæ
sit media proportionalis inter AK & AC, ideoq; in dimidiata ratione resistentiæ) incrementum resistentiæ data temporis particu-
la factum per lineolam KL, & contemporaneum velocitatis incrementum per lineolam PQ; & centro C Asymptotis rectangulis
CA, CH describatur Hyperbola quævis BNS, erectis perpendicu-
lis AB, KN, LO, PR, QS occurrens in B, N, O, R, S.   Quo-
niam AK est ut APq., erit hujus momentum KL ut illius momentum $2APQ$, id est ut AP in KC.   Nam velocitatis incrementum PQ, per motus Leg. 2. proportionale est vi generanti
KC.   Componatur ratio ipsius KL cum ratione ipsius KN, &
fiet rectangulum $KL \times KN$ ut $AP \times KC \times KN$; hoc est, ob da-
tum rectangulum $KC \times KN$, ut AP.   Atqui areæ Hyperbolicæ

Leibniz's annotations on his recently discovered copy of Newton's *Principia* in-
dicate he had already discovered the calculus independently.   Reprinted with
permision from E.A. Fellmann, *The Principia and continental mathematicians*,
Notes Rec. R. Soc. Lond. 42 (1988), 13-34.

Part I

The Derivative

1

Instantaneous Velocity and the Derivative

A car speeds down the highway for three hours (see Figure 1.1.). Suppose that the distance traveled after t hours is given by the formula

$$\text{mileage} = f(t) = 20t^2.$$

After 1 hour, the car has traveled $f(1) = 20(1)^2 = 20$ miles; after 2 hours, 80 miles; after 3 hours, 180 miles. The car is speeding up! Its average speed for the first hour is 20 mph; for the second, $80 - 20 = 60$ mph; for the third, 100 mph. How fast is it going at the end? Is it possible to derive a formula for its velocity at every instant? This is a hard question. There is no obvious way to do it. We will find a way, and we will call it differential calculus.

The hard thing about computing the velocity at an instant is that the distance traveled and the elapsed time are both zero, so that you cannot just divide. The trick is to compute the average velocity over shorter and shorter time periods of length Δt. The symbol Δ is the

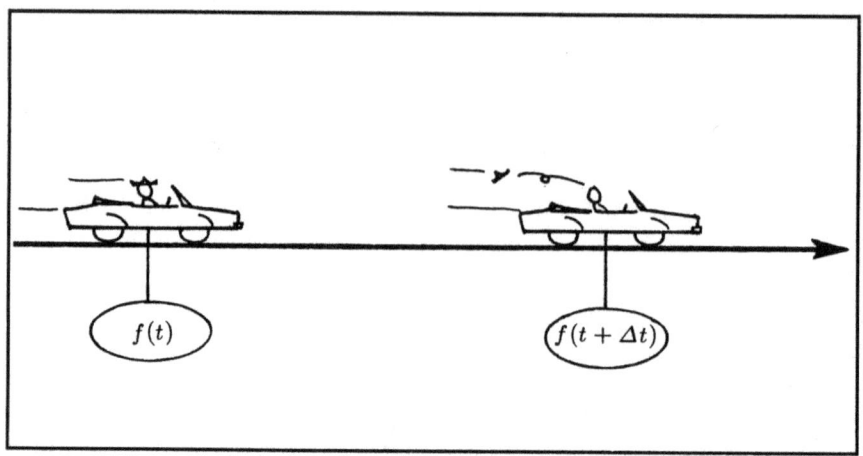

Figure 1.1. If $f(t)$ gives a car's odometer reading at time t, the distance traveled between time t and $t + \Delta t$ is $f(t + \Delta t) - f(t)$.

capital Greek letter delta (see table at end of book). It is used to denote "the change in." Thus Δt ("delta t") denotes the change in t.

The distance traveled between time t and $t + \Delta t$ is

$$
\begin{aligned}
f(t + \Delta t) - f(t) &= 20(t + \Delta t)^2 - 20t^2 \\
&= 20[t^2 + 2t\Delta t + \Delta t^2 - t^2] \\
&= 20[2t\Delta t + \Delta t^2].
\end{aligned}
$$

(Be sure you know the formula $(a + b)^2 = a^2 + 2ab + b^2$ by heart.)

Dividing this distance by Δt, we see that the average velocity over this short time period is $20[2t + \Delta t]$. As Δt gets smaller, we approach the instantaneous velocity $20[2t] = 40t$. In particular, after three hours the car is going 120 mph.

Thus from the formula for the mileage $f(t) = 20t^2$, we derived another formula for the speed $f'(t) = 40t$. This new function f' is called the *derivative* of the first function f.

The key part of the process was computing the instantaneous velocity as a limiting ratio as Δt goes to 0. If you took Δt to be 0 to start with, the distance would also be 0, and the velocity $0/0$ would be undefined.

Finding derivatives is half of calculus. (Finding "integrals" will be the other half.)

Mathematically, the process of deriving the speed function from the mileage function can be summarized by the following:

1.1. Theoretical Definition of the Derivative

The derivative $f'(t)$ of a function $f(t)$ is the function defined by considering ratios

$$\frac{f(t + \Delta t) - f(t)}{\Delta t}$$

and seeing what limiting value you get as Δt goes to 0:

$$f'(t) = \lim_{\Delta t \to 0} \frac{f(t + \Delta t) - f(t)}{\Delta t}.$$

If f represents mileage, f' represents speed. If f is any function, f' gives its rate of change.

The following rules let you compute lots of derivatives almost effortlessly.

1.2. Power Rule

For any positive integer power $f(t) = t^n$, $f'(t) = nt^{n-1}$.

Warning. This applies only to powers of t, not to $(t^2 + 1)^n$, for example.

1.3. Multiplier Rule

The derivative of a constant multiple is just that multiple of the derivative: $(af)' = af'$.

Example 1.1. Using these these two rules, we can recover our earlier computation that $(20t^2)' = 40t$ without any work:

$$
\begin{aligned}
(20t^2)' &= 20(t^2)' \quad \text{by Rule 1.3} \\
&= 20 \cdot 2t^1 \quad \text{by Rule 1.2} \\
&= 40t.
\end{aligned}
$$

Proof of the Power Rule

Case $n = 1, f(t) = t$.

$$f'(t) = \lim_{\Delta t \to 0} \frac{f(t + \Delta t) - f(t)}{\Delta t} = \lim_{\Delta t \to 0} \frac{t + \Delta t - t}{\Delta t}$$

$$= \lim_{\Delta t \to 0} 1 = 1t^0.$$

Case $n = 2, f(t) = t^2$.

$$f'(t) = \lim_{\Delta t \to 0} \frac{f(t + \Delta t) - f(t)}{\Delta t} = \lim_{\Delta t \to 0} \frac{(t + \Delta t)^2 - t^2}{\Delta t}$$

$$= \lim_{\Delta t \to 0} \frac{t^2 + 2t\Delta t + (\Delta t)^2 - t^2}{\Delta t} = \lim_{\Delta t \to 0} (2t + \Delta t) = 2t.$$

Case $n = 3, f(t) = t^3$.
Here we will need the formula $(a + b)^3 = a^3 + 3a^2 b + 3ab^2 + b^3$.

$$f'(t) = \lim_{\Delta t \to 0} \frac{f(t + \Delta t) - f(t)}{\Delta t} = \lim_{\Delta t \to 0} \frac{(t + \Delta t)^3 - t^3}{\Delta t}$$

$$= \lim_{\Delta t \to 0} \frac{t^3 + 3t^2 \Delta t + 3t(\Delta t)^2 + (\Delta t)^3 - t^3}{\Delta t}$$

$$= \lim_{\Delta t \to 0} (3t^2 + 3t\Delta t + (\Delta t)^2) = 3t^2.$$

General Case $f(t) = t^n$.

$$f'(t) = \lim_{\Delta t \to 0} \frac{f(t + \Delta t) - f(t)}{\Delta t} = \frac{(t + \Delta t)^n - t^n}{\Delta t}$$

$$= \lim_{\Delta t \to 0} \frac{t^n + nt^{n-1}\Delta t + (\text{something})(\Delta t)^2 + \ldots - t^n}{\Delta t}$$

$$= \lim_{\Delta t \to 0} (nt^{n-1} + (\text{something})\Delta t + \ldots)$$

$$= nt^{n-1}.$$

Proof of the Multiplier Rule

$$\begin{aligned}
(af)'(t) &= \lim_{\Delta t \to 0} \frac{af(t + \Delta t) - af(t)}{\Delta t} \\
&= a \lim_{\Delta t \to 0} \frac{f(t + \Delta t) - f(t)}{\Delta t} \\
&= af'(t).
\end{aligned}$$

There is another rule you may think obvious:

1.4. Sums Rule

The derivative of a sum of two functions is just the sum of the derivatives: $(f + g)' = f' + g'$.

Example 1.2. Let $f(t) = 3t^3 - 7t^2$. Then

$$\begin{aligned}
f'(t) &= (3t^3)' + (-7t^2)' && \text{by Sums Rule} \\
&= 3(t^3)' + (-7)(t^2)' && \text{by Multiplier Rule} \\
&= 3(3t^2) + (-7)(2t) && \text{by Power Rule} \\
&= 9t^2 - 14t.
\end{aligned}$$

The derivative of a constant is 0. For example, if $f(t) = 7$, then $f'(t) = 0$. This can be proved from the definition of the derivative or viewed as a special case of the Power Rule:

$$(7)' = (7t^0)' = 7(t^0)' = (7)(0)t^{-1} = 0.$$

Derivatives with respect to other quantities. So far we have used derivatives for rates of change with respect to time, t. We could study equally well atmospheric pressure p as a function $p(h)$ of height h. Perhaps near the ground

$$p(h) = 100 - .01h.$$

Then the rate of change of the pressure $p'(h)$ as height increases is given by

$$p'(h) = -.01.$$

Near the ground, as you rise pressure decreases at a rather small rate.

The notation $y = \mathbf{f}(\mathbf{x})$. It is most common to use x instead of t or h for the variable and to use y for the value of $f(x)$.

Example 1.3. Suppose

$$y = f(x) = -7x^{10} + 5x^3 + 2.$$

Then

$$y' = -70x^9 + 15x^2.$$

Notation. Sometimes one abbreviates Δf for $f(x + \Delta x) - f(x)$. In this notation,

$$f'(x) = \lim_{\Delta x \to 0} \frac{\Delta f}{\Delta x}.$$

Leibniz's notation

$$f'(x) = \frac{df}{dx}$$

well expresses the idea of the derivative as the ratio of an infinitesimal change df in f to the infinitesimal change dx in x which causes it.

Exercises 1

Compute the derivative $f'(t)$ of the given function $f(t)$. (If $f(t)$ represents mileage, then $f'(t)$ represents speed.)

1. $f(t) = t^7 + t^5$

2. $f(t) = t^{100} + t^{50} + t^{10}$

3. $f(t) = -6t^3 + 12t^2 - 4t + 7$

4. $f(t) = -6t^3 + 12t^2 - 4t + 9$

5. $f(t) = \frac{1}{10}t^{10} + \frac{1}{9}t^9 + \frac{1}{8}t^8$

6. $f(t) = \frac{1}{3}t^3 + c,$ c is a constant

7. $f(t) = (3t + 5)^2$

   (Expand first: if you try to use the Power Rule immediately, you will get the wrong answer.)

8. $f(t) = (-5t^2 + t - 1)^2$ (Expand first.)

9. $f(t) = (-2t^2 + 1)^3$ (Expand first.)

10. $f(t) = (t^2 + 1)^6$ (Expand first.)

11. $f(t) = (3t^2 - 7t + 1)(t^2 + t - 1)$ (Expand first.)

12. $f(t) = (\frac{1}{3}t^3 + \frac{1}{2}t^2 + t + 1)(\frac{1}{5}t^5 + \frac{1}{4}t^4 + t + 1)$ (Expand first.)

Use the definition of the derivative to compute the derivatives of the following functions.

13. $f(t) = 3t + 5$

14. $f(t) = -7t - 4$

15. $f(t) = t^2 + t + 1$

16. a. $f(t) = 6t^2 - t - 8$
    b. $f(t) = 7t^3$

Given y, compute the derivative $y' = \dfrac{dy}{dx}$.

17. $y = 5x^2 + 7x + 9$

18. $y = 100x^2 + \dfrac{1}{100}x$

Use the definition of the derivative to compute the derivatives of the following functions.

19. $y = \dfrac{1}{2}x^2$

20. $y = x^4$

2

Geometric Interpretation of the Derivative as the Slope of the Graph

As you heard in algebra, the graph of $y = mx + b$ in the x-y plane is a line with slope m and y-intercept b; see Figure 2.1. The slope is $\frac{\Delta y}{\Delta x}$, the rate of change of y with respect to x, which for the line is the constant m. Our new-found calculus gives us the same answer:

$$\frac{dy}{dx} = m(x^1)' + b' = m \cdot 1 + 0 = m.$$

Note that the slope of a line is constant, the same at all points.

The graph of $y = x^2$ is a parabola (Figure 2.2). Here the slope is variable and harder to compute. By calculus, the slope is given by $y' = 2x$. When $x = 0$, the slope is 0. When $x = -1$, the slope is -2. When $x = 2$, the slope is 4. These slopes are easier to see by drawing the tangent lines which have the same slope.

For general functions $y = f(x)$, the derivative function y' or $f'(x)$ can be interpreted as giving the slope of the graph at the point (x, y).

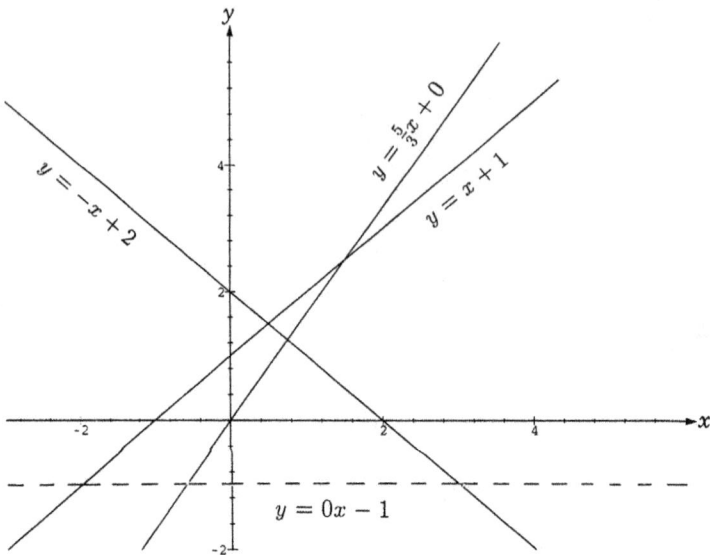

Figure 2.1. The graph of $y = mx + b$ in the x-y plane is a line with slope m and y-intercept b.

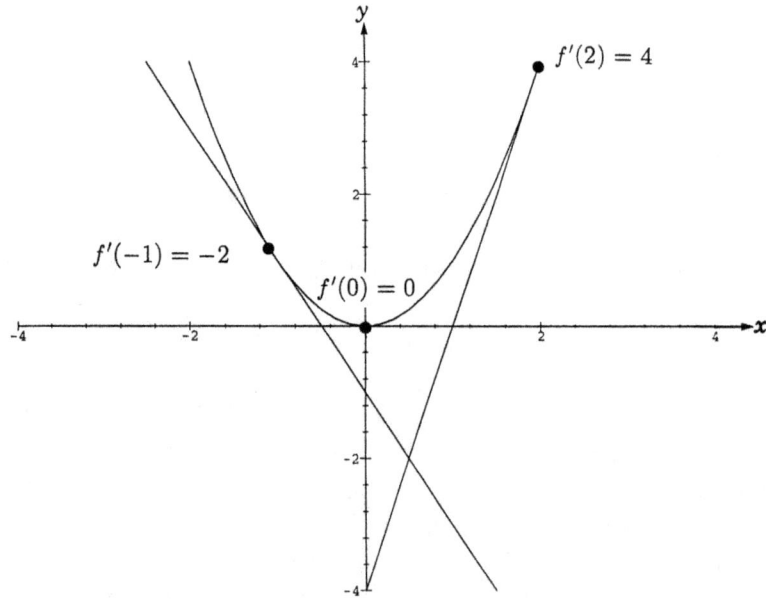

Figure 2.2. The graph of $y = x^2$ is a parabola, with slope given by $y' = 2x$. Tangent lines have the same slope as the graph.

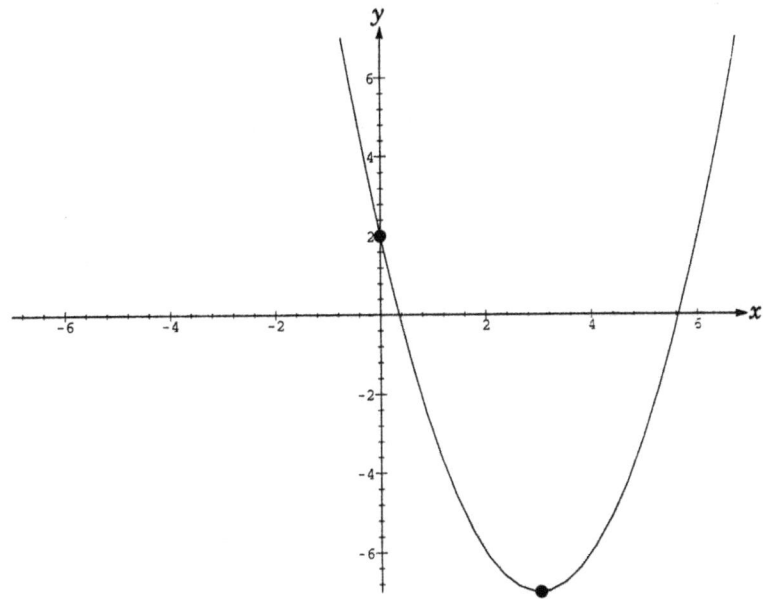

Figure 2.3. The slope of the graph of $y = f(x) = x^2 - 6x + 2$ is given by $y' = 2x - 6$. It is 0 when $x = 3$.

Example 2.1. Find a formula for the slope of the graph of $y = f(x) = x^2 - 6x + 2$. Where is the slope 0?

Solution. The slope is given by $y' = 2x - 6$. It is 0 when $x = 3$; see Figure 2.3.

Example 2.2. Find the equation for the line tangent to the graph of $y = f(x) = x^3 - 5x^2 + 6$ at the point $(2, -6)$.

Solution. Since $f'(x) = 3x^2 - 10x$, the slope of the graph at $(2, -6)$ is $f'(2) = -8$. Since the tangent line has the same slope, its formula is

$$y = mx + b = -8x + b.$$

To solve for b, plug in the point $(2, -6)$:

$$-6 = (-8)(2) + b \Rightarrow b = 10.$$

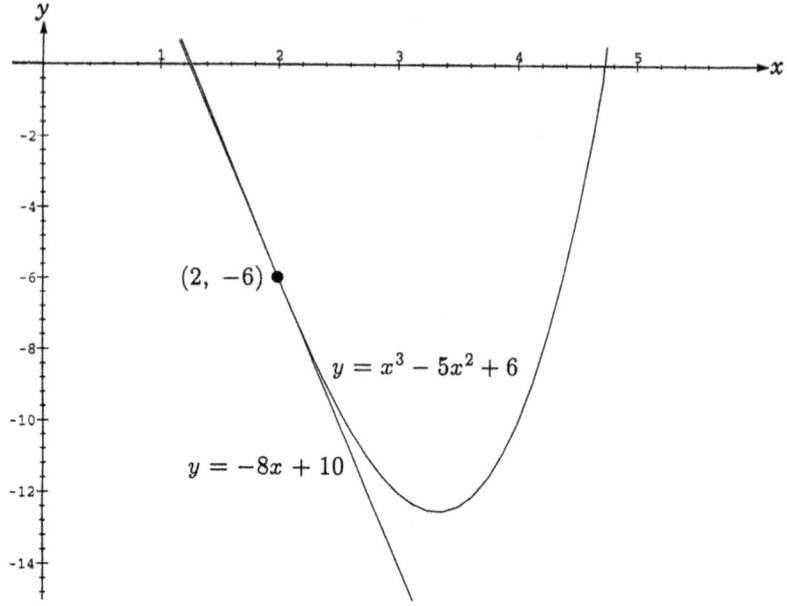

Figure 2.4. The line tangent to the graph of $y = f(x) = x^3 - 5x^2 + 6$ at the point $(2, -6)$ is given by $y = -8x + 10$.

Therefore the equation for the tangent line is

$$y = -8x + 10.$$

(See Figure 2.4.)

Exercises 2

First, sketch the graph of the following functions for $-3 \le x \le 3$ by plotting $y = f(x)$ at $x = -3, -2, -1, 0, 1, 2, 3$. Second, label each point with a guess for the slope at that point. Finally, label each point with the actual slope, computed as

$$\frac{dy}{dx} = y' = f'(x).$$

1. $y = x^2 - 4$

2. $y = \dfrac{1}{2}x^2 - x - 4$

3. $y = -\dfrac{x^2}{3} + \dfrac{x}{3} + 2$

4. $y = \dfrac{1}{3}x^3$

5. $y = x^3 - 9x$

6. $y = -\dfrac{x^3}{3} + \dfrac{x^2}{3} + x - 2$

7. Find the equation of the line tangent to the graph of $y = x^2 - 4$ at the point $(2, 0)$.

8. Find the equation of the line tangent to the graph of $y = x^3 - 3x$ at the point $(2, 2)$.

9. Find the equation of the line tangent to the graph of $y = x^5$ at the point where $x = 1$.

10. Find the equation of the line tangent to the graph of $y = (x^2 + 1)^2$ at the point where $x = -2$.

3

The Product and Quotient Rules

This section begins with a rule for the derivative of a product of functions, such as

$$y = (5x^4 + 6x^3 - x^2 + 7x - 2)(x^3 + 2x^2 + x - 7), \qquad (3.1)$$

which will spare us the necessity of multiplying them out first.

3.1. Product Rule

$$(fg)' = f'g + fg'.$$

Note that the derivative of a product is not just the product of the derivatives; if it were, $(x^2)' = (x \cdot x)'$ would be $1 \cdot 1 = 1$ instead of $2x$. Notice that our Product Rule says $(x \cdot x)' = 1 \cdot x + x \cdot 1 = 2x$, the right answer.

Here is the reason for the unexpected formula. A small change Δx leads to small changes Δf and Δg. The new value of the product,

$$(f + \Delta f)(g + \Delta g) = fg + (\Delta f)g + f(\Delta g) + (\Delta f)(\Delta g),$$

has the crossterms $(\Delta f)g$ and $f(\Delta g)$ as well as the expected $(\Delta f)(\Delta g)$. These crossterms are the important ones, because if (Δf) and (Δg) are both small, their product $(\Delta f)(\Delta g)$ is negligible. In greater detail, the change $\Delta(fg)$ in the product is obtained by subtracting the original value fg from the new value $(f + \Delta f)(g + \Delta g)$:

$$\Delta(fg) = (f + \Delta f)(g + \Delta g) - fg = (\Delta f)g + f(\Delta g) + (\Delta f)(\Delta g).$$

Hence

$$
\begin{aligned}
(fg)' &= \lim_{\Delta x \to 0} \frac{\Delta(fg)}{\Delta x} = \lim_{\Delta x \to 0} \left(\frac{\Delta f}{\Delta x} g + f \frac{\Delta g}{\Delta x} + \frac{\Delta f}{\Delta x} \frac{\Delta g}{\Delta x} \Delta x \right) \\
&= f'g + fg' + f'g' \cdot 0 \\
&= f'g + fg'.
\end{aligned}
$$

Example 3.1. Find y' for y as defined by Equation 3.1.

Solution. Here $y = f(x)g(x)$, with $f(x) = 5x^4 + 6x^3 - x^2 + 7x - 2$ and $g(x) = x^3 + 2x^2 + x - 7$. Therefore

$$
\begin{aligned}
y' &= f'(x)g(x) + f(x)g'(x) \\
&= (20x^3 + 18x^2 - 2x + 7)(x^3 + 2x^2 + x - 7) \\
&\quad + (5x^4 + 6x^3 - x^2 + 7x - 2)(3x^2 + 4x + 1).
\end{aligned}
$$

This answer may be checked against what you get by multiplying out $f(x)$ times $g(x)$, and then differentiating.

The rule for the derivative of a quotient is a little more complicated.

3.2. Quotient Rule

$$\left(\frac{f}{g} \right)' = \frac{f'g - fg'}{g^2}.$$

Example 3.2. If

$$y = \frac{x^3 - 2x + 7}{x^4 + 7x + 1},$$

then

$$y' = \frac{(3x^2 - 2)(x^4 + 7x + 1) - (x^3 - 2x + 7)(4x^3 + 7)}{(x^4 + 7x + 1)^2}$$

$$= \frac{-x^6 + 6x^4 - 14x^3 + 3x^2 - 51}{(x^4 + 7x + 1)^2}.$$

Proof. First note that a change in the quotient

$$\Delta\left(\frac{f}{g}\right) = \frac{f + \Delta f}{g + \Delta g} - \frac{f}{g} = \frac{fg + (\Delta f)g - fg - f\Delta g}{g(g + \Delta g)}$$

$$= \frac{(\Delta f)g - f\Delta g}{g(g + \Delta g)}.$$

Therefore

$$\left(\frac{f}{g}\right)' = \lim_{\Delta x \to 0} \frac{\Delta\left(\frac{f}{g}\right)}{\Delta x} = \lim_{\Delta x \to 0} \frac{\frac{\Delta f}{\Delta x}g - f\frac{\Delta g}{\Delta x}}{g\left(g + \frac{\Delta g}{\Delta x}\Delta x\right)}$$

$$= \frac{f'g - fg'}{g(g + g' \cdot 0)} = \frac{f'g - fg'}{g^2}.$$

Exercises 3

Compute the derivatives y'.

1. $y = (x^2 + 1)(x^3 + 1)$

   (Two ways: Product Rule and expanding first.)

2. $y = (3x^7 + 5x)(x^2 - 8x + 4)$ (Two ways.)

3. $y = (x - 1)(x^5 + x^4 + x^3 + x^2 + x + 1)$ (Two ways.)

4. $y = \left(\frac{1}{4}x^4 + \frac{1}{3}x^3 + \frac{1}{2}x^2 + x\right)(4x^4 + 3x^3 - 2x^2 + x)$ (Two ways.)

5. $y = \dfrac{5x + 6}{x^2 + 1}$

6. $y = \dfrac{3x^2 - 2x + 5}{4x^2 + 5x - 7}$

7. $y = \dfrac{x^3 - x}{x^5 - x^3 + 1}$

8. $y = \dfrac{\dfrac{1}{2}x^2 - x}{\dfrac{1}{3}x^3 - x + 1}$

9. $y = \dfrac{x^2}{2}$

10. $y = \dfrac{x^2 + 9x + 2}{7}$

11. $y = (x^3 + 1)(x^2 + 1)(x + 1)$ (Two ways.)

12. $y = (5x^7 - x^5 - 3x^3 + 7)(x^2 + 2x + 3)(x^2 - 2x - 3)$ (Two ways.)

13. $y = \dfrac{(x^3 - 2x^2 + 7x + 1)(x^2 - x + 3)}{x^3 + 1}$

14. $y = \left(\dfrac{x^2 + 1}{x^3 + 1}\right)\left(\dfrac{x^4 + 1}{x^5 + 1}\right)$

(Use Product Rule. For each factor, use Quotient Rule.)

15. $y = \dfrac{(x^2 + 1)(x^4 + 1)}{(x^3 + 1)(x^5 + 1)}$

(Use Quotient Rule. For numerator and for denominator, use Product Rule.)

16. Exercise 1.11 by Product Rule.

17. Exercise 1.12 by Product Rule.

18. Exercise 1.7 as a product.

19. If $f = g^2$, prove $f'(1) = 2g(1)g'(1)$.

4

The Chain Rule and Implicit Differentiation

The Chain Rule gives a rule for differentiating a "chain" of nested or composite functions, such as $y = (x^2 + 1)^{10}$, which will spare us the necessity of multiplying them out first. Here y is the tenth power of some thing, namely $x^2 + 1$; if we call that thing u, then we can write $y = u^{10}$, where $u = x^2 + 1$.

4.1. Chain Rule

If $y = f(u)$, and $u = g(x)$, then

$$\frac{dy}{dx} = \frac{dy}{du} \cdot \frac{du}{dx}.$$

For $y = (x^2 + 1)^{10} = u^{10}$ with $u = x^2 + 1$,

$$\frac{dy}{dx} = \frac{dy}{du} \cdot \frac{du}{dx}$$

$$= 10u^9 \cdot 2x$$

$$= 10(x^2+1)^9(2x)$$

$$= 20x(x^2+1)^9.$$

The idea is that if an airplane goes 5 times as fast as a train and the train goes 4 times as fast as a moped, then the plane goes $5 \cdot 4 = 20$ times as fast as the moped. That is,

$$\Delta y \approx \frac{dy}{du}\Delta u \,,$$

and

$$\Delta u \approx \frac{du}{dx}\Delta x \,,$$

so that

$$\Delta y \approx \frac{dy}{du}\frac{du}{dx}\Delta x.$$

In the limit,

$$\frac{dy}{dx} = \lim \frac{\Delta y}{\Delta x} \quad \text{is exactly} \quad \frac{dy}{du}\frac{du}{dx}.$$

We will not give a careful proof.

The Chain Rule is easy to remember in Leibniz's notation, where it *looks* as if we are just canceling the du's, although officially du/dx is a single symbol that we are not supposed to take apart.

Example 4.1. Find y' if $y = ((x^2+1)^3+4)^5$.

Solution. We need to use the Chain rule three times in succession, starting with $(u^5)' = 5u^4$ times u' .

$$
\begin{aligned}
y' &= 5((x^2+1)^3+4)^4 \text{ times } ((x^2+1)^3+4)' \\
&= 5((x^2+1)^3+4)^4 \text{ times } 3(x^2+1)^2 \text{ times } (x^2+1)' \\
&= 5((x^2+1)^3+4)^4 \text{ times } 3(x^2+1)^2 \text{ times } 2x.
\end{aligned}
$$

Usually one simplifies by putting the simplest factors in front, to obtain

$$y' = 30x(x^2+1)^2((x^2+1)^3+4)^4.$$

4.2. Implicit Differentiation

The Chain Rule will allow us to solve for y' without solving for y first. By the Chain Rule, the derivative of u^5 is $(u^5)' = 5u^4 u'$. Applying that last equation to y itself yields $(y^5)' = 5y^4 y'$.

Example 4.2. If $y^5 + xy + x^2 = 3$, find y'.

Solution. Differentiating the whole equation yields
$$5y^4 y' + (1 \cdot y + xy') + 2x = 0.$$

Note how the Product Rule was used to differentiate xy. Solving for y' yields
$$y' = \frac{-2x - y}{5y^4 + x}.$$

Note. It is important to know the main independent variable, often x, with respect to which we are taking the derivative. For example $(x^2)' = 2x$, while $(y^2)' = 2yy'$. If instead y were the independent variable, x were a function $x(y)$ of y, and we wanted to compute derivatives with respect to y, then $(y^2)' = 2y$ and $(x^2)' = 2x\,x'$.

Exercises 4

Find $\dfrac{dy}{dx}$.

1. $y = (x^2 + 2x + 3)^2$

   (Check by multiplying out and then differentiating.)

2. $y = (x^3 - 4)^6$

   (I dare you to check by multiplying out and then differentiating.)

3. $y = \left(\dfrac{x^3}{3} + 1\right)^5 + \left(\dfrac{x^2}{2} + 1\right)^4 + 3$

4. $y = \left(5x^3 + x - 3\right)^4 + \left(5x^3 - x + 3\right)^4$

5. $y = \left((x^2 + 1)^{10} + 1\right)^8$

6. $y = \left((x^4 + 1)^4 + (x^3 + 1)^3 + (x^2 + 1)^2\right)^5$

7. $y = \dfrac{\left((x^3 + 7)^4 + x\right)^5}{x^2 + 1}$

8. $x^2 + y^2 = 1$

9. $x^3 + y^3 = 1$

10. $y^4 + xy + x^4 = 3$

11. $y^4 + x^2 y^2 + x^4 = 3$

12. $x^2 = y$

13. $x^3 = y^5$

14. $\left((x^2 + y^2)^4 + (x^4 + y^4)^2\right)^{10} + x^5 + y^5 = 100$

15. $y = \left(\dfrac{x}{x + 1}\right)^{10}$

16. $y = \dfrac{x^{10}}{(x + 1)^{10}}$

17. Exercise 1.7 using Chain Rule.

18. Exercise 1.8 using Chain Rule.

19. Exercise 1.9 using Chain Rule.

20. Exercise 1.10 using Chain Rule.

21. Find $y'(1)$ if $\quad y = \dfrac{1}{\dfrac{1}{x^2+1} + \dfrac{1}{x^2+2} + \dfrac{1}{x^2+3}}$.

22. Suppose $y = \dfrac{u^2}{v^3}$. Find an expression for y' in terms of $u, u', v,$ and v'.

5

The Extended Power Rule

The Power Rule actually holds for any real power, including negative powers and fractional powers.

5.1. Extended Power Rule

For any real number r ,

$$(x^r)' = rx^{r-1} \qquad (x > 0).$$

Hence by the Chain Rule,

$$(u^r)' = ru^{r-1}u'.$$

Examples 5.1.

$$(\sqrt{x})' = (x^{\frac{1}{2}})' = \frac{1}{2}x^{-\frac{1}{2}} = \frac{1}{2\sqrt{x}}. \qquad (5.1)$$

$$\left(\frac{1}{x}\right)' = (x^{-1})' = -x^{-2} = -\frac{1}{x^2}. \qquad (5.2)$$

$$(x^\pi)' = \pi x^{\pi-1}. \tag{5.3}$$

$$
\begin{aligned}
\left(\sqrt{6x^2+5x+3}\,\right)' &= \left((6x^2+5x+3)^{\frac{1}{2}}\right)' \tag{5.4}\\
&= \frac{1}{2}(6x^2+5x+3)^{-\frac{1}{2}}(6x^2+5x+3)'\\
&= \frac{1}{2}(6x^2+5x+3)^{-\frac{1}{2}}(12x+5)\\
&= \frac{12x+5}{2\sqrt{6x^2+5x+3}}\,.
\end{aligned}
$$

Warning. The power must be a constant.

$$
\begin{aligned}
(2^x)' &\neq x\cdot 2^{x-1}.\\
(x^x)' &\neq x\cdot x^{x-1}.
\end{aligned}
$$

We do not know how to differentiate these functions yet. We'll come to the first in Chapter 9.

5.2. Powers

Recall from algebra that

$$
\begin{aligned}
x^{\frac{1}{2}} &= \sqrt{x} & (x>0)\\
x^{\frac{1}{3}} &= \sqrt[3]{x}\\
x^{\frac{1}{q}} &= \sqrt[q]{x} & (x\geq 0 \quad \text{if } q \text{ is even})\\
x^{\frac{p}{q}} &= \sqrt[q]{x^p} & (x\geq 0 \quad \text{if } q \text{ is even})\\
x^{-r} &= \frac{1}{x^r}\\
(x^r)^s &= x^{rs}\\
x^r x^s &= x^{r+s}
\end{aligned}
$$

but $x^r + x^s$ and $x^r y^s$ cannot be simplified.

Having defined rational powers

$$x^{\frac{p}{q}} = \sqrt[q]{x^p},$$

an irrational power such as $x^{\sqrt{2}}$ could be defined as a limit using better and better decimal approximations of $\sqrt{2}$:

$$x^{\sqrt{2}} \;\; = \;\; \lim x^{1.4}, x^{1.41}, x^{1.414}, \ldots, \quad \text{where}$$

$$x^{1.4} \;\; = \;\; x^{\frac{7}{5}} = \sqrt[5]{x^7},$$

$$x^{1.41} \;\; = \;\; x^{\frac{141}{100}} = \sqrt[100]{x^{141}}, \quad \text{etc.}$$

To pursue this approach faithfully, one must now prove that x^r so defined has the right properties: *e.g.* $\left(x^{\sqrt{2}}\right)^3 = x^{3\sqrt{2}}$. As you might guess, it does turn out all right.

5.3. Derivation of the Extended Power Rule

The discussion of the validity of the Extended Power Rule falls into three cases.

Case 1. The power r is a positive rational number $r = p/q$ in lowest terms. Since $y = x^{p/q}$, then $y^q = x^p$. Implicit differentiation yields $qy^{q-1}y' = px^{p-1}$. Hence

$$y' \;\; = \;\; \frac{p}{q}\frac{x^{p-1}}{y^{q-1}} = \frac{p}{q}\frac{x^p}{y^q}\frac{y}{x} = \frac{p}{q}\frac{y}{x},$$

because $x^p = y^q$. Therefore,

$$y' = \frac{p}{q}\frac{x^{p/q}}{x'} \;\; = \;\; \frac{p}{q}x^{\frac{p}{q}-1} = rx^{r-1}$$

as desired.

Case 2. The power r is a positive irrational number, such as $\sqrt{2}$. In this case we assume that x is positive. Recall we define $x^{\sqrt{2}}$ as the limit of $x^{1.4}, x^{1.41}, x^{1.414}, \ldots$. A similar approach can be used for any r. Let $q_1, q_2, q_3, \ldots$ be the decimal approximations, and define $x^r = \lim x^{q_i}$. To compute the derivative, one would like to say that

$$
\begin{aligned}
\left(x^{\sqrt{2}}\right)' &= \left(\lim x^{1.4}, x^{1.41}, x^{1.414}, \ldots\right)' \\
&= \lim 1.4x^{.4}, 1.41x^{.41}, 1.414x^{.414}, \ldots \\
&= \sqrt{2}x^{\sqrt{2}-1}.
\end{aligned}
$$

The second equality uses rules too advanced for our presentation here, but it all turns out all right in this case. Many calculus books postpone irrational powers until after logs and exponentials.

Case 3. The power r is negative: $r = -s$. In this case we usually assume that x is positive. Now we just use the Quotient Rule.

$$
\begin{aligned}
(x^r)' &= \left(\frac{1}{x^s}\right)' = \frac{0 - sx^{s-1}}{x^{2s}} \\
&= -sx^{-s-1} = rx^{r-1}.
\end{aligned}
$$

Example 5.2. If $y = \sqrt{x^2 + 2x + 5}$, find y'.

Solution.

$$
\begin{aligned}
y &= \left(x^2 + 2x + 5\right)^{\frac{1}{2}} \\
y' &= \frac{1}{2}\left(x^2 + 2x + 5\right)^{-\frac{1}{2}}\left(x^2 + 2x + 5\right)'
\end{aligned}
$$

by the Chain Rule

$$
\begin{aligned}
&= \frac{1}{2}\left(x^2 + 2x + 5\right)^{-\frac{1}{2}}(2x + 2) \\
&= \frac{x + 1}{\sqrt{x^2 + 2x + 5}}.
\end{aligned}
$$

Exercises 5

Simplify or say "cannot be simplified."

1. a. $64^{\frac{1}{2}}$, $64^{\frac{1}{3}}$, $64^{-\frac{1}{6}}$

   b. $9^{\frac{3}{2}}$, $\left(\frac{1}{4}\right)^{\frac{3}{2}}$, $\left(\frac{9}{4}\right)^{\frac{3}{2}}$, $\left(\frac{9}{4}\right)^{-\frac{3}{2}}$

   c. $\sqrt[3]{x^9}$, $\sqrt{x^2 + y^2}$, $(\sqrt{x})^4$, $x^{\frac{1}{2}} x^{-\frac{5}{2}}$, $\dfrac{1}{\left(x^{-\frac{1}{2}}\right)^2}$

   d. $(x^2)^{\frac{2}{3}}$, $(x^3 y^6)^{\frac{1}{3}}$, $(x^3 + y^3)^{\frac{3}{4}}$, $x^2 + x^4$.

Find the derivative y':

2. $y = 3x^{\frac{1}{3}} + x^{\frac{2}{3}}$

3. $y = x^{\frac{3}{4}} + x^{\frac{4}{3}}$ $\qquad$ $(x > 0)$

4. $y = \sqrt{x} + \sqrt[3]{x}$ $\qquad$ $(x > 0)$

5. $y = \dfrac{1}{\sqrt{x}}$ $\qquad$ $(x > 0)$

6. $y = \sqrt{x^2 + 1}$

7. $y = \sqrt[3]{2x - 3}$

8. $y = \dfrac{\sqrt{x} + 1}{\sqrt{x} + 1}$ $\qquad$ $(x > 0)$

9. $y = \sqrt{\sqrt{x + 1} + 2}$ $\qquad$ $(x > -1)$

10. $y = \dfrac{1}{\sqrt[3]{x\sqrt{x} + 4x - \sqrt{x} + 4}}$ $\qquad (x > 0)$

11. $y = \sqrt{1 + \sqrt{1 + \sqrt{x}}}$ $\qquad (x > 0)$

6

Sines, Cosines, and
Their Derivatives

As you travel around the circle of Figure 6.1 counterclockwise, the angle θ with the x-axis increases from $0°$ to $360°$, or from 0 radians to 2π radians. The conversion factor is $360°/2\pi$ radians. In calculus, everything turns out much simpler if you use radians instead of degrees. For example, the length of an arc of a unit circle equals its measure in radians: $s = \theta$. For a circle of radius r, the length $s = r\theta$.

As you travel around the unit circle and θ increases from 0 to 2π, the height or y-coordinate is called the sine of θ: $y = \sin\theta$. The x-coordinate is called the cosine of θ: $x = \cos\theta$. Figure 6.2 shows $(x, y) = (\cos\theta, \sin\theta)$ at important angles, best memorized at once.

First notice the angles: 2π or $360°$ take you all the way around; π or $180°$ halfway; $\pi/2$ or $90°$ to the y-axis; $\pi/4$ or $45°$ halfway to the y-axis; $\pi/6$ or $30°$ a third of the way to the y-axis; $\pi/3$ or $60°$ two-thirds of the way to the y-axis. Next notice that there are just a few special values of the cosines and sines:

$$0, \ \pm\frac{1}{2}, \ \pm\frac{1}{\sqrt{2}}, \ \pm\frac{\sqrt{3}}{2}, \ \pm 1.$$

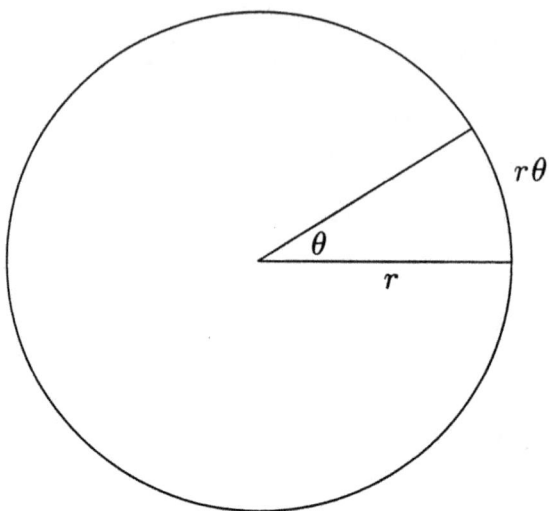

Figure 6.1. The length s of an arc of θ radians in a circle of radius r satisfies $s = r\theta$.

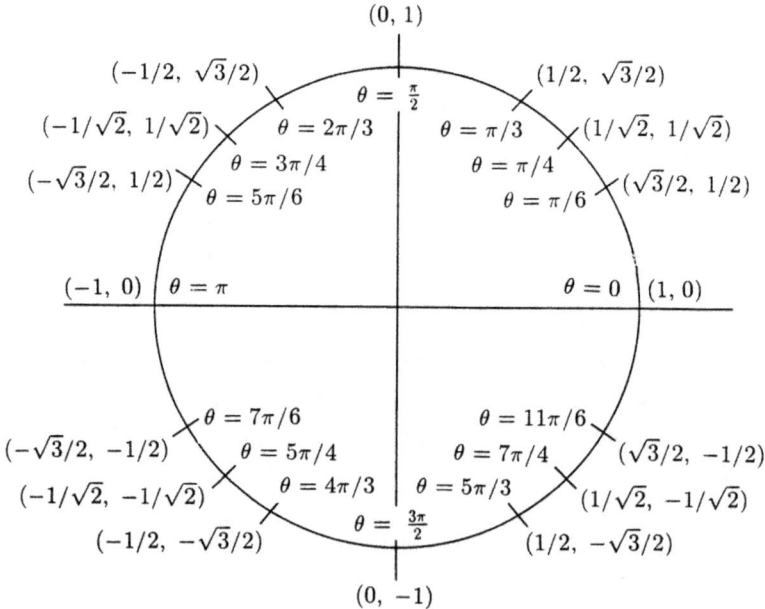

Figure 6.2. The values $(\cos\theta, \sin\theta)$ at important angles.

Except for the obvious values of 0 and 1, $\sqrt{3}/2$ is the big one, and $1/2$ is the small one. Finally notice that in the first quadrant where $0 < \theta < \pi/2$, the sine and cosine are both positive. In the second quadrant where $\pi/2 < \theta < \pi$, the cosine is negative but the sine is still positive. In the third quadrant where $\pi < \theta < 3\pi/2$, the sine and cosine are both negative. Finally in the fourth quadrant where $3\pi/2 < \theta < 2\pi$, the cosine is positive but the sine is still negative.

A Few Important Trigonometric Identities

1. $\sin^2\theta + \cos^2\theta = 1$

2. $\sin(A + B) = \sin A \cos B + \cos A \sin B$

   e.g., $\sin(2\theta) = 2\sin\theta\cos\theta$

3. $\cos(A + B) = \cos A \cos B - \sin A \sin B$

   e.g., $\cos(2\theta) = \cos^2\theta - \sin^2\theta$

Note that $\sin^2\theta$ means $(\sin\theta)^2$.

If instead of $x = \cos\theta$ and $y = \sin\theta$, you graph $y = \sin x$ and $y = \cos x$, then instead of the unit circle you get the curves of Figure 6.3.

For a right triangle as in Figure 6.4, $\sin\theta = $ opposite/hypotenuse, $\cos\theta = $ adjacent/hypotenuse, and $\tan\theta = $ opposite/adjacent.

Of course you will need to know the derivatives of these important functions. Fortunately, up to sign, each turns out to be the derivative of the other.

6.1. Sine and Cosine Rules

$$(\sin x)' = \cos x; \quad (\cos x)' = -\sin x.$$

Hence by the Chain Rule,

$$(\sin u)' = (\cos u)u'; \quad (\cos u)' = -(\sin u)u'.$$

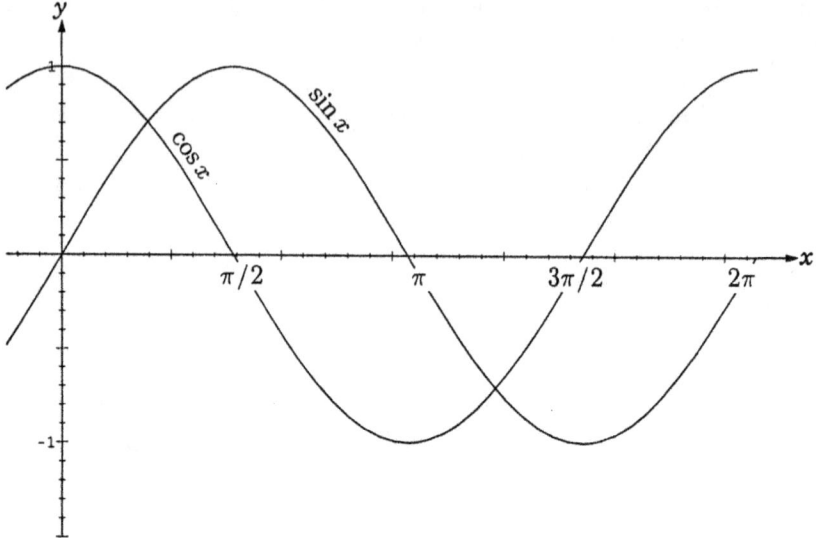

Figure 6.3. The graphs of the functions $y = \sin x$ and $y = \cos x$.

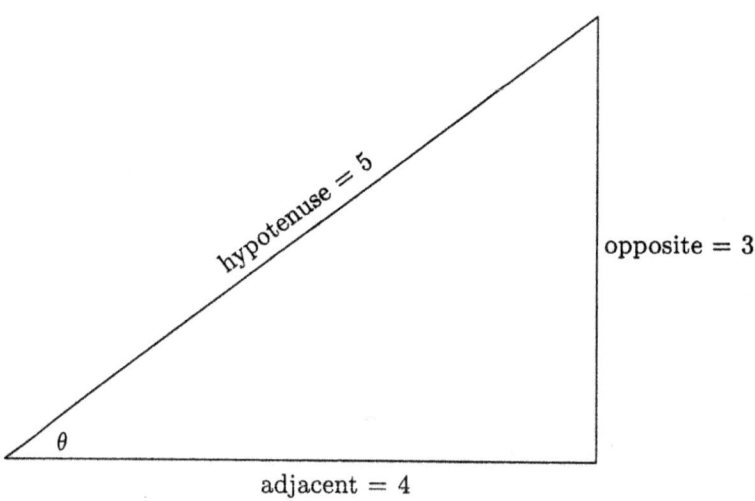

Figure 6.4. $\sin\theta = \frac{\text{opposite}}{\text{hypotenuse}} = \frac{3}{5}$, $\cos\theta = \frac{\text{adjacent}}{\text{hypotenuse}} = \frac{4}{5}$, $\tan\theta = \frac{\text{opposite}}{\text{adjacent}} = \frac{3}{4}$.

Examples 6.1

If $f(x) = 4\cos x$, then $f'(x) = -4\sin x$.

If $f(x) = \cos 4x$, then by the Chain Rule

$$f'(x) = (-\sin 4x)(4x)' = -4\sin 4x.$$

Proof of the Sine and Cosine Rules

Michael Livshits has shown me the following short derivation of the Sine and Cosine Rules, which Alan Durfee points out was given by C. S. Ogilvy in *The American Mathematical Monthly* (Vol. 67, 1960, p. 673). Figure 6.5 shows a large right triangle with angle θ, height $\sin\theta$, and width $\cos\theta$. The smaller right triangle has hypotenuse approximately equal to the change $\Delta\theta$ in θ. (The exact $\Delta\theta$ is a tiny arc of the circle, well approximated by the hypotenuse of the small triangle.) Its height represents the increase in $\sin\theta$, and its width

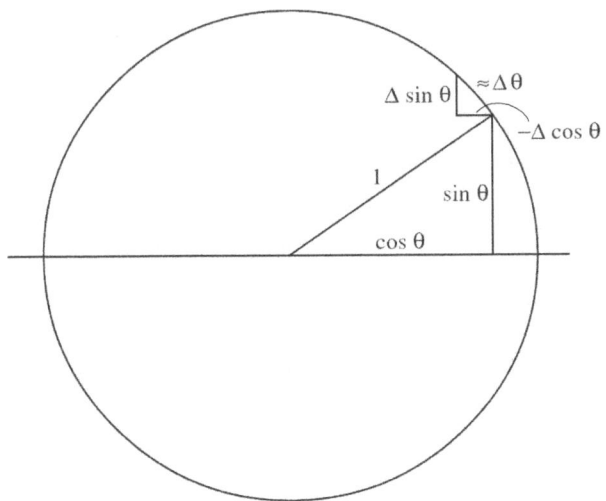

Figure 6.5. By similar triangles, $\dfrac{\Delta\sin\theta}{\Delta\theta} \approx \dfrac{\cos\theta}{1}$, and $\dfrac{\Delta\cos\theta}{\Delta\theta} \approx -\dfrac{\sin\theta}{1}$.

represents the decrease in $\cos\theta$. The little triangle is the same shape as the big one, just rotated 90 degrees. Hence it is a similar triangle,

$$\frac{\Delta\sin\theta}{\Delta\theta} \approx \frac{\cos\theta}{1},$$

and

$$\frac{\Delta\cos\theta}{\Delta\theta} \approx -\frac{\sin\theta}{1}.$$

In the limit as θ approaches 0, the ratios on the left yield the derivatives, the approximations become exact, and

$$(\sin\theta)' = \cos\theta; \quad (\cos\theta)' = -\sin\theta.$$

We have considered only angles θ between 0 and $\pi/2$, but a similar argument applies to any angle.

Exercises 6

Find the derivative $\dfrac{dy}{dx}$.

1. $y = 5\sin x$

2. $y = -\dfrac{3}{2}\cos x$; also, what is $y'(\dfrac{\pi}{3})$?

3. $y = -4\sin 5x - 6\cos\dfrac{1}{2}x$; also, what is $y'(\dfrac{\pi}{2})$?

4. $y = \dfrac{1}{\sqrt{2}}\sin\sqrt{2}x - \cos\pi x$

5. $y = \cos x^2$ (This means $\cos(x^2)$.)

6. $y = \cos(x^2 - 3x + 7)$

7. $y = \cos^2 x$ (This means $(\cos x)^2$.)

8. $y = \cos^2 x - 3\cos x + 7$;   also, what is $y'(\frac{7\pi}{6})$?

9. $y = \sin^5 x - 9\sin x$;   also, what is $y'(\frac{3\pi}{4})$?

10. $y = \sqrt[3]{\sin x} + \sqrt[5]{\cos 5x}$

11. $y = \cos \sqrt{x^2 + 7}$

12. $y = \sin \sqrt[3]{x^3 - 2x + 1}$

13. $y = \dfrac{\sin x + \cos x}{\sin x - \cos x}$

14. $y = \dfrac{\cos^2 x + 3\sin x}{\sin^2 x + x^2}$

15. $y = \dfrac{1}{\sqrt{\sin^2 3x + 7}}$

16. $y = \pi\sin^3 \pi x$;   also, what is $y'(\frac{5}{3})$?

17. To differentiate, say which methods you would use in what order:

Power Rule	Extended Power Rule	Product Rule
Quotient Rule	Chain Rule	Simplification

(a) $\sqrt{x^2 + 1}$

(b) $\dfrac{\sqrt{x}}{x}$

(c) $\dfrac{\cos 3x + \sin 3x}{x^4 + 3x^2 + 1}$

(d) $(x^2 + 1)^5 (x^2 - 1)^{10}$

(e) $\dfrac{1}{\sqrt{t+1}}$

(f) $\sqrt{\dfrac{11}{x^3}}$

(g) $\cos^4 4x^4$

(h) $(g(x))^3$

(i) $g(x^3)$

(j) $x^3 g(x)$

18. Find y' if $y = \cos^4(5x - \frac{\pi}{2})$.

7

Maxima and Minima

Maxima-minima problems are what life is all about. We humans want to maximize profit or happiness; we want to minimize labor, expense, pollution. We spend our lives looking for maximums and minimums, or to use the official plurals, *maxima* and *minima*.

Most maxima-minima problems have *extreme* answers. The way to minimize expenses today is not to spend any money. The way to minimize pollution is not to pollute.

For other maxima-minima problems there is a *critical* balance between competing factors. What size warehouse should Sears build to maximize profit? If the warehouse is too big, it will cost too much to build and maintain. If the warehouse is too small, supplies frequently will run out and business will be lost.

Summary. Maxima-minima problems can have two kinds of answers:

1. extreme cases

2. critical cases.

In general you want to maximize some quantity y which depends upon another quantity x, which you can vary within certain limits, as in the following.

Puzzle. For which x between given extremes of 0 and 1 is

$$y = x - 2x^3 \quad (0 \leq x \leq 1)$$

biggest (maximum)? Try to guess if you can. When x is 0, y is 0; when x is 1, y is $1 - 2 = -1$. Maybe 0 is the biggest value for y, or maybe y is positive somewhere in between 0 and 1. Try $x = 1/2$. When x is 1/2, y is 1/4; yes, it is positive. Is that the biggest? When x is .4, y is .272, still bigger! Table 7.1 gives more values, graphed in Figure 7.1.

Table 7.1. Some values of $y = x - 2x^3$ for $0 \leq x \leq 1$. Where is y biggest?

x	0	.1	.3	.4	.5	.7	.8	1.0
$y = x - 2x^3$	0	.098	.246	.272	.250	.014	-.224	-1.000

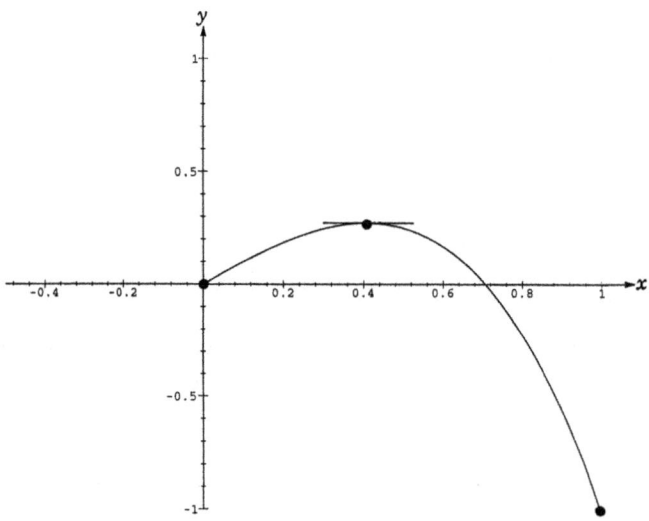

Figure 7.1. The graph of $y = x - 2x^3$ for $0 \leq x \leq 1$. Where y is biggest, the slope $y' = 0$.

So $x = .4$, $y = .27200$ looks pretty good, but when x is .41, y is .27216, still a little better. Is there any way to find the exact critical spot where y is biggest? *Yes, it is a place where the derivative or slope $y' = 0$.* (See Figure 7.1.) In this case, since $y = x - 2x^3$, therefore $y' = 1 - 6x^2$, and $y' = 0$ when $1 = 6x^2$, when $x = 1/\sqrt{6} \approx .40825$. The maximum value is exactly

$$y = \frac{1}{\sqrt{6}} - 2(\frac{1}{\sqrt{6}})^3 = \frac{1}{\sqrt{6}} - \frac{1}{3\sqrt{6}} = \frac{2}{3\sqrt{6}} \approx .27217.$$

This procedure of checking the extreme and critical cases works in general:

7.1. Maxima-Minima

To find the maxima or minima of a function $y = f(x)$, *look at*

1. *extreme cases (sometimes given, sometimes* $x \to \pm\infty$, *sometimes y undefined, perhaps because some denominator is 0), and*

2. *critical cases, where* $y' = 0$ *or is undefined.*

Figures 7.2–7.4 illustrate different cases that can occur. Generally there are only a few extreme and critical cases to check.

The following examples illustrate the method of 7.1 for finding maxima or minima. We begin by reworking the above puzzle methodically and finding the minimum as well as the maximum.

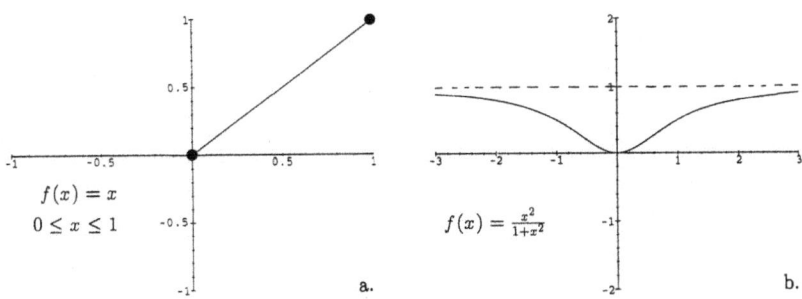

Figure 7.2. *Extreme cases.* a. On the interval $[0,1]$, $f(x) = x$ attains its maximum at the right endpoint. b. As $x \to \pm\infty$, $f(x) = \frac{x^2}{1+x^2}$ grows towards 1, but it never reaches a maximum.

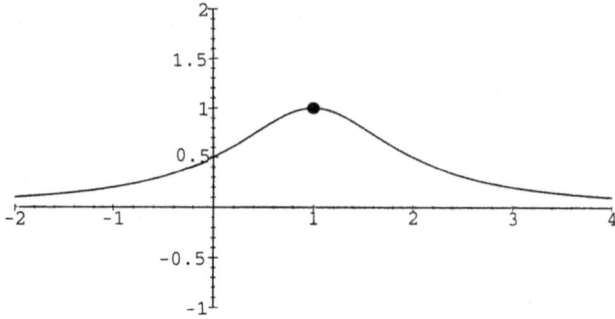

Figure 7.3. Note that y can have a maximum value when $y' = 0$.

Example 7.1. Find the maxima and minima of

$$y = x - 2x^3 \quad \text{for } 0 \le x \le 1.$$

Solution. First check the extreme cases. When $x = 0$, $y = 0$; when $x = 1$, $y = -1$. Second check the critical cases. Here $y' = 1 - 6x^2 = 0$ only when $x = 1/\sqrt{6} \approx .40825$ and

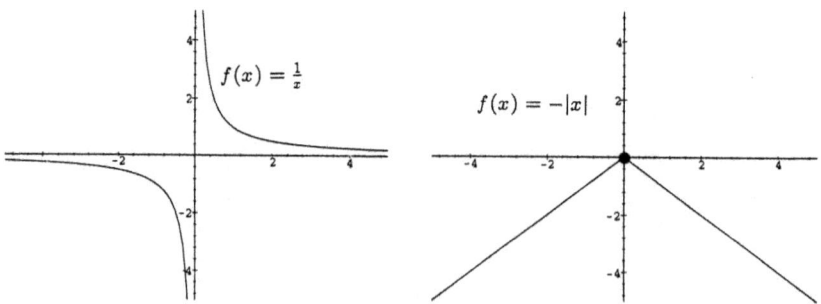

Figure 7.4. Near a point where y or y' is undefined, y can get big. Since $y = 1/x$ is undefined at 0, it never reaches a maximum. For $y = -|x|$, y' is undefined at the maximum (see Chapt. 13).

$$y = \frac{1}{\sqrt{6}} - 2\left(\frac{1}{\sqrt{6}}\right)^3$$

$$= \frac{1}{\sqrt{6}} - \frac{1}{3\sqrt{6}}$$

$$= \frac{2}{3\sqrt{6}} \approx .27217.$$

As is often the case, there is no problem with y or y' being undefined. Comparing the three values of y, we find the maximum is $\frac{2}{3\sqrt{6}}$ at $x = \frac{1}{\sqrt{6}}$ and the minimum is -1 at $x = 1$:

x	0	$\frac{1}{\sqrt{6}}$	1
$y = x - 2x^3$	0	$\frac{2}{3\sqrt{6}}$	-1
		$\approx .27217$	
		maximum	minimum

We call $y = \frac{2}{3\sqrt{6}}$ the *maximum value* and $x = \frac{1}{\sqrt{6}}$ the *maximum point*; we call $y = -1$ the *minimum value* and $x = 1$ the *minimum point*.

Example 7.2. Find the maxima and minima of

$$y = x^{\frac{2}{3}} \quad \text{for} \quad x \geq -1.$$

Solution. In the first extreme case, when $x = -1$, $y = \sqrt[3]{(-1)^2} = 1$. There is no given upper limit on x, so we consider what happens as x gets huge: y gets huge. In other words, as $x \to +\infty, y \to +\infty$. Therefore y never actually reaches a maximum (it just keeps getting bigger).

For the critical cases: here $y' = \frac{2}{3}x^{-\frac{1}{3}}$ is never 0, but it is undefined at $x = 0$, where $y = 0$. Comparing these three cases, we conclude that y has a minimum of 0 at $x = 0$ and no maximum (except perhaps at $+\infty$, which does not count). No matter how big y gets, it gets still bigger as x gets still bigger.

Example 7.3. Find the maxima and minima of

$$y = |x| = \begin{cases} x & \text{if } x > 0 \\ -x & \text{if } x < 0 \\ 0 & \text{if } x = 0 \,. \end{cases}$$

Solution. The extreme cases occur as $x \to \pm\infty$, when $y \to +\infty$. Hence y has no maximum. For the critical cases, note that

$$y' = \begin{cases} 1 & \text{if } x > 0 \\ -1 & \text{if } x < 0 \\ \text{undefined} & \text{if } x = 0 \,. \end{cases}$$

There is no place where $y' = 0$, and y' is undefined at 0, where $y = 0$. Comparing these cases, we see that y has a minimum of 0 at 0, and no maximum.

Warning. It is not always easy ahead of time to know whether the maxima and minima will be extreme or critical cases. You have to consider both possibilities.

The following is a typical sort of theorem you might find in a mathematics book.

7.2. Theorem

A continuous function $y = f(x)$ on a closed interval $[a, b]$ attains a maximum and a minimum.

At first it sounds right and obvious that a function will have a maximum and a minimum somewhere, and the details about "continuous" and "on a closed interval $[a, b]$" do not sound important.

The "closed interval $[a, b]$" is just the set of all x's between two extremes a and b, inclusive:

$$[a, b] = \{x : a \le x \le b\}.$$

Yet the theorem fails on the "open interval (a, b)," which does not include a and b:

$$(a, b) = \{x : a < x < b\}.$$

For example, on the open interval $(0, 1)$, the function $y = x$ never reaches the maximum value of 1, because x is not allowed to be 1 (see Figure 7.5). Similarly, on the open interval $(0, 1)$, the function $y = \frac{1}{x}$ never reaches a maximum: it just gets bigger and bigger as x gets close to 0.

Moreover, the theorem fails if f is not continuous, as illustrated by the function on $[0,1]$ graphed in Figure 7.6, which jumps down "discontinuously" just when it is about to reach its maximum. For more on continuity, see Chapter 13.

So this is a very limited theorem! Indeed, as we have already seen, many maxima-minima problems do not have solutions. Problems without ideal solutions occur in real life, too. For example, you might want to minimize the weight of an adequately strong airplane wing. One way to reduce weight without reducing strength involves putting lots of small holes in the wing: the tinier and more numerous

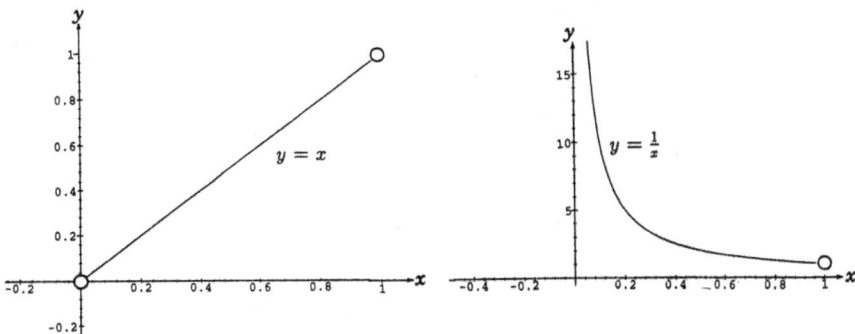

Figure 7.5. On the *open* interval $(0,1)$, the function $y = x$ never reaches the maximum value of 1, because x is not allowed to be 1. Similarly, the function $y = 1/x$ never reaches a maximum: it just gets bigger and bigger as x gets close to 0.

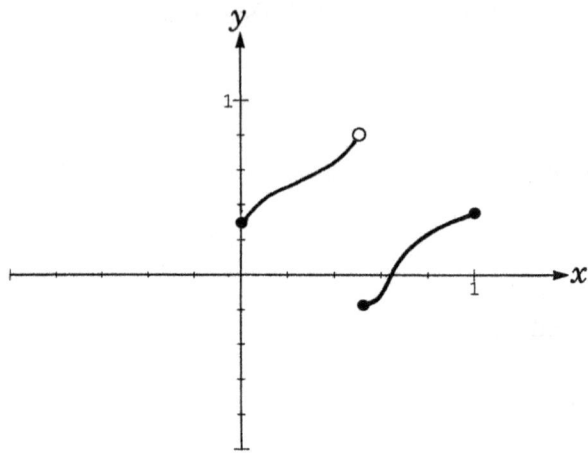

Figure 7.6. Even on the closed interval [0,1], a function may fail to reach a maximum by jumping down "discontinuously" just when it is about to reach its maximum.

the holes, the lighter the wing. There is no minimum weight: you cannot make it all holes!

The proof of this delicate theorem depends on a deep understanding of the real numbers, continuous functions, and the difference between closed and open intervals. You can learn about it in a course on *real analysis*, the theory behind calculus.

Examples 7.2 and 7.3 illustrated the need to understand how functions grow and behave as $x \to \infty$.

7.3. Rates of Growth

As x gets large, x^5 gets much bigger than x^2. We say that "x^5 grows faster than x^2," or "x^2 grows slower than x^5," and we write

$$x^2 \ll x^5.$$

Technically this just means that

$$\lim_{x \to +\infty} \frac{x^2}{x^5} = \lim_{x \to +\infty} \frac{1}{x^3} = 0;$$

i.e., that x^5 surpasses any multiple of x^2. Note that even

$$100x^2 \ll \frac{1}{100}x^5,$$

because

$$\lim_{x \to +\infty} \frac{100x^2}{\frac{1}{100}x^5} = \lim_{x \to +\infty} \frac{10000}{x^3} = 0.$$

Similarly $\sqrt{x} = x^{\frac{1}{2}} \ll x$. On the other hand, $5x^7 \not\ll 100x^7$, because

$$\lim_{x \to +\infty} \frac{5x^7}{100x^7} = \frac{1}{20} \neq 0.$$

In summary,

$$\sqrt[3]{x} \ll \sqrt{x} \ll x \ll x^2 \ll x^{100}.$$

7.4. Limits

Certain limits are easy to compute because one or two terms overwhelm the others, which can therefore be ignored.

Example 7.4.

$$\lim_{x \to \infty} \frac{3x^5 + x^3 + \sqrt{x}}{4x^5 + x^2 + \sqrt{x}} = \lim_{x \to \infty} \frac{3x^5}{4x^5} = \frac{3}{4}.$$

(The other terms are overwhelmed by x^5. This can be proved by dividing numerator and denominator by x^5 and watching the numerator approach 3 and the denominator approach 4.)

Another important class of limits

$$\lim_{x \to a} \frac{N(x)}{D(x)}$$

involves a fraction which blows up because the denominator $D(x)$ goes to zero, while the numerator $N(x)$ stays away from zero. In this case the limit may be $+\infty$ or $-\infty$, and it sometimes depends on whether $x > a$ or $x < a$. The "right-hand limit" $\lim_{x \to a^+}$ assumes that $x > a$. The "left-hand limit" $\lim_{x \to a^-}$ assumes that $x < a$.

Example 7.5.

$$\lim_{x \to +\infty} 2x^5 - 100x^4 = +\infty;$$

$$\lim_{x \to 3^+} \frac{1}{x - 3} = +\infty, \qquad \text{while}$$

$$\lim_{x \to 3^-} \frac{1}{x - 3} = -\infty,$$

because when $x > 3$, $x - 3$ is positive and hence the fraction is positive, whereas when $x < 3$, $x - 3$ is negative and hence the fraction is negative.

Example 7.6.

$$\lim_{x \to 4^+} \frac{x(x - 4)(x + 5)}{(x - 4)^3} = \lim_{x \to 4^+} \frac{x(x + 5)}{(x - 4)^2} = +\infty$$

because the numerator and denominator are both positive. In this case, also

$$\lim_{x \to 4^-} \frac{x(x - 4)(x + 5)}{(x - 4)^3} = \lim_{x \to 4^-} \frac{x(x + 5)}{(x - 4)^2} = +\infty,$$

and we can simply say that

$$\lim_{x \to 4} \frac{x(x - 4)(x + 5)}{(x - 4)^3} = \lim_{x \to 4} \frac{x(x + 5)}{(x - 4)^2} = +\infty.$$

Note that to compute this limit it was important to factor the numerator and denominator first.

Example 7.7.

$$\lim_{x \to -1^+} \frac{x^2 + 4x + 4}{x^2 - 5x - 6} = \lim_{x \to -1^+} \frac{(x + 2)^2}{(x - 6)(x + 1)} = -\infty$$

because the numerator is positive, $(x - 6)$ is negative, and $(x + 1)$ is positive, while

$$\lim_{x \to -1^-} \frac{x^2 + 4x + 4}{x^2 - 5x - 6} = \lim_{x \to -1^-} \frac{(x + 2)^2}{(x - 6)(x + 1)} = +\infty.$$

7.5. A Hard Maxima-Minima Problem

Find the maxima and minima of

$$f(x) = \frac{x^2 + x + 1}{x^2 - 2x - 2}.$$

First we compute the derivative by the Quotient Rule:

$$
\begin{aligned}
f'(x) &= \frac{(2x+1)(x^2 - 2x - 2) - (x^2 + x + 1)(2x - 2)}{(x^2 - 2x - 2)^2} \\
&= \frac{(2x^3 - 4x^2 - 4x + x^2 - 2x - 2) - (2x^3 - 2x^2 + 2x^2 - 2x + 2x - 2)}{(x^2 - 2x - 2)^2} \\
&= \frac{-3x^2 - 6x}{(x^2 - 2x - 2)^2} \\
&= \frac{-3x(x+2)}{(x^2 - 2x - 2)^2}.
\end{aligned}
$$

The derivative is 0 whenever the numerator is 0 (and the denominator is not 0), *i.e.*, when x is 0 or -2. By the original formula we compute that $f(0) = -1/2$, $f(-2) = 3/6 = 1/2$. These are candidates for the minimum and maximum.

For the extreme cases,

$$\lim_{x \to \pm\infty} f(x) = 1;$$

i.e., $f(x) \to 1$ without reaching it.

Finally, f and f' will be undefined if the denominator $x^2 - 2x - 2 = 0$. First we try to factor this expression

$$x^2 - 2x - 2 = (x-?)(x-?),$$

but we fail. Then we resort to the quadratic formula

$$x = \frac{2 \pm \sqrt{4+8}}{2} = 1 \pm \frac{\sqrt{4 \cdot 3}}{2} = 1 \pm \sqrt{3}.$$

These are real roots, the denominator does go to 0, and the fraction blows up. Indeed,

$$f(x) = \frac{x^2 + x + 1}{(x - (1 + \sqrt{3}))(x - (1 - \sqrt{3}))}.$$

As $x \to 1 + \sqrt{3}$ from the right, the numerator is positive, the first factor in the denominator is a very small positive, and the second factor in the denominator is about $2\sqrt{3} > 0$, so that $f(x) \to +\infty$. Similarly as $x \to 1 + \sqrt{3}$ from the left, $f(x) \to -\infty$, because the first factor in the denominator becomes negative. These limiting values of $\pm\infty$ tell us that $f(x)$ can never reach a maximum or minimum. We do not even have to check $1 - \sqrt{3}$. In fact, if we had started with this part of the analysis, we would not have had to do anything else. Thus, $f(x)$ has no maxima and no minima.

Exercises 7

Find the maximum and minimum values and points.

1. $f(x) = x^2 - 3x - 4$ $0 \leq x \leq 2$

2. $f(x) = x^3 - 12x$ $0 \leq x \leq 3$

3. $f(x) = \sqrt{x} - \sqrt[3]{x}$ $0 \leq x \leq 1$

4. $f(x) = x^2 - x^3$ $0 \leq x \leq 1$

5. $f(x) = x^2 - 5x + 3$ (Unless specified otherwise, always assume the largest possible domain, here $-\infty < x < +\infty$.)

6. $f(x) = x^4 - 4x$

7. $f(x) = x^3 - x$

8. $f(x) = x^{20} - x^{10}$

9. Give examples or sketches if possible of

     a. a function on $[0, 1]$ with a maximum but no minimum;

     b. a continuous function on $(0, 1)$ with a maximum but no minimum;

c. a continuous function on $[0, 1]$ with a maximum but no minimum.

10. Arrange the following functions in a chain of $\ll$.

$$\frac{1}{100}x^5, \ \frac{1}{100}x^{10.1}, \ \frac{1}{100}\sqrt{x}, \ x^{10}, \ \sqrt[3]{x}, \ 100x^4$$

Compute the following limits.

11. a. $\lim\limits_{r \to 0} \dfrac{1}{r}$ $\qquad\qquad$ $(r > 0)$

$\quad$ b. $\lim\limits_{r \to 0} \left(2\pi r^2 + \dfrac{2000}{r}\right)$ $\qquad\qquad$ $(r > 0)$

$\quad$ c. $\lim\limits_{r \to \infty} \left(2\pi r^2 + \dfrac{2000}{r}\right)$ $\qquad\qquad$ $(r > 0)$

12. a. $\lim\limits_{x \to \infty} \dfrac{x}{x^2 + 1}$

$\quad$ b. $\lim\limits_{x \to \infty} \dfrac{x^2}{x^2 + 1}$

$\quad$ c. $\lim\limits_{x \to \infty} \dfrac{x^4}{x^2 + 1}$

13. $\lim\limits_{x \to \infty} \dfrac{3x^2 + 2x + 1}{4x^2 + 10x + 7}$

14. $\lim\limits_{x \to \infty} \dfrac{5x + 2x^2 + 7x^3}{\sqrt[3]{x} + x^3}$

15. $\lim\limits_{x \to \infty} \dfrac{5x + 2x^{10}}{x\sqrt{x} + x^{20}}$

16. $\lim\limits_{x \to \infty} \dfrac{\sqrt{x^{10} + x}}{x^5}$

17. $\displaystyle\lim_{x\to\infty} \frac{.01x^4}{10^6 x^3}$

Find the maximum and mininum values and points.

18. $f(x) = x^2 + 16x^{-1}$ $\qquad\qquad$ $(x > 0)$

19. $f(r) = 2\pi r^2 + \dfrac{4000}{r}$ $\qquad$ $(r > 0)$

20. $f(x) = \dfrac{x^2}{1 + x^2}$

21. $f(x) = \dfrac{x^4}{1 + x^2}$

22. $\quad$ a. $f(x) = \dfrac{x^2}{1 + x^4}$

$\qquad$ b. $f(x) = \dfrac{1 + x^2}{2 + x^2}$

23. $\quad$ a. $f(x) = 2x^3 + 3x^2 - 6x + 1$ $\quad (x \geq 0)$

$\qquad$ b. $f(x) = 2x^3 + 3x^2 + 6x + 1$ $\quad (x \geq 0)$

24. $f(x) = \sin x$

(You should just know this one without doing any work.)

25. $f(x) = \cos x$

26. $f(x) = \left(\dfrac{\sin x}{x}\right)^2,$ $\quad x \neq 0$

27. $f(x) = \sin x - \cos x$

Compute the following limits.

28. a. $\lim\limits_{x \to 8+} \dfrac{5}{x - 8}$

    b. $\lim\limits_{x \to 8-} \dfrac{5}{x - 8}$

29. a. $\lim\limits_{x \to -4+} \dfrac{(x - 3)(x - 4)}{(x + 3)(x + 4)}$

    b. $\lim\limits_{x \to -4-} \dfrac{(x - 3)(x - 4)}{(x + 3)(x + 4)}$

    c. $\lim\limits_{x \to -4+} \dfrac{(x + 3)(x + 4)}{(x - 3)(x - 4)}$

30. a. $\lim\limits_{x \to 3+} \dfrac{x^2 + 5x - 14}{x^2 + x - 12}$

    b. $\lim\limits_{x \to 3-} \dfrac{x^2 + 5x - 14}{x^2 + x - 12}$

31. a. $\lim\limits_{x \to -6+} \dfrac{x^2 + 3x + 2}{x^2 + 9x + 18}$

    b. $\lim\limits_{x \to -6-} \dfrac{x^2 + 3x + 2}{x^2 + 9x + 18}$

32. a. $\lim\limits_{x \to 5+} \dfrac{9x^2}{x^2 - 10x + 25}$

    b. $\lim\limits_{x \to 5-} \dfrac{9x^2}{x^2 - 10x + 25}$

    c. $\lim\limits_{x \to -3+} \dfrac{x^2 + 7x + 10}{x^2 - 10x + 11}$

33.   a. $\displaystyle\lim_{x\to 1^+} \frac{x^2 + 2x + 5}{x^2 - 3x + 2}$

    b. $\displaystyle\lim_{x\to 1^-} \frac{x^2 + 2x + 5}{x^2 - 3x + 2}$

    c. $\displaystyle\lim_{x\to 2^+} \frac{x^2 + 2x + 5}{x^2 - 3x + 2}$

    d. $\displaystyle\lim_{x\to 2^-} \frac{x^2 + 2x + 5}{x^2 - 3x + 2}$

    e. $\displaystyle\lim_{x\to 3^+} \frac{x^2 + 2x + 5}{x^2 - 3x + 2}$

    f. $\displaystyle\lim_{x\to 3^-} \frac{x^2 + 2x + 5}{x^2 - 3x + 2}$

Find the maximum and minimum values and points.

34. $f(x) = \dfrac{1}{1 - x^2} \qquad x \neq \pm 1$

35. $f(x) = \dfrac{x}{1 - x^2} \qquad x \neq \pm 1$

* 36. $f(x) = \dfrac{\cos x}{1 + x^2}$

* 37. $f(x) = \dfrac{x^2 + 2x + 5}{x^2 - 2x + 2}$

38. $f(x) = \dfrac{x^2 + 2x + 5}{x^2 - 3x + 2}$

39. $f(x) = -x^{\frac{4}{3}}$

40. $f(x) = 8x^{\frac{2}{3}} - x^{\frac{4}{3}}$

41. The temperature T in degrees centigrade on the orbiting space shuttle from noon until 8:00 p.m. is given by the formula

$$T(t) = t^3 - 9t^2 + 15t,$$

where t is the time in hours since noon. What are the highest and lowest temperatures during that period?

8

Maxima-Minima
Real-World Problems

This chapter will consider the more interesting kind of maxima-minima problems that occur in real life. Reasoning through a real-world minimization problem requires the following steps:

8.1. Procedure for Solving Maxima-Minima Real-World Problems

1. Identify the quantity y you are trying to maximize or minimize (such as cost), the variable x you can adjust to minimize y (such as the radius of some cylinder), and any prescribed quantity (such as volume).

2. Get a formula for y as a function of x: $y = f(x)$. This involves putting everything in terms of x.

3. As in the previous section, look at extreme cases and critical cases (or where y or y' is undefined). In most real-world problems, the maxima and minima are extreme cases, often

obviously; in others, the maxima and minima are obviously not extreme cases.

8.2. The Garden Problem

You have an opportunity to work in your neighbor's garden for up to four hours Saturday at $6 per hour. How long should you work to earn the most money?

Solution. Here the answer is obviously the extreme case of working all four hours. For practice we will now see how the above three-step procedure leads to this answer. The quantity y to be maximized is the earnings; the variable x is the number of hours to work. The formula for y in terms of x is

$$y = f(x) = 6x.$$

In the extreme cases, when $x = 0$, $y = 0$, and when $x = 4$, $y = 24$. Since $y' = 6$ is never 0, there are no critical cases. The maximum value of $24 occurs at $x = 4$. The minimum value of $0 occurs at $x = 0$.

8.3. The Rectangular Pen Problem

Find the shape of the biggest rectangular pen for your pet that you can make with 36 feet of fencing. See Figure 8.1.

Solution. The quantity to be maximized is the area A. For the variable, we could use the length ℓ or the width w; then the unused variable is determined by the prescribed fence perimeter $2\ell + 2w = 36$. Let's use ℓ as the variable. Then $w = (36 - 2\ell)/2 = 18 - \ell$ and

$$A = \ell w = \ell(18 - \ell) = 18\ell - \ell^2.$$

For this problem the extreme cases of long, narrow pens (see Figure 8.2) are obviously no good, so we turn to the critical cases:

$$0 = A' = 18 - 2\ell = 2(9 - \ell).$$

Hence

$$\ell = 9, \quad w = 18 - \ell = 9.$$

The maximum area is given by the 9×9 square, with area 81.

Figure 8.1. Find the shape of the biggest rectangular pen for your pet that you can make with 36 feet of fencing.

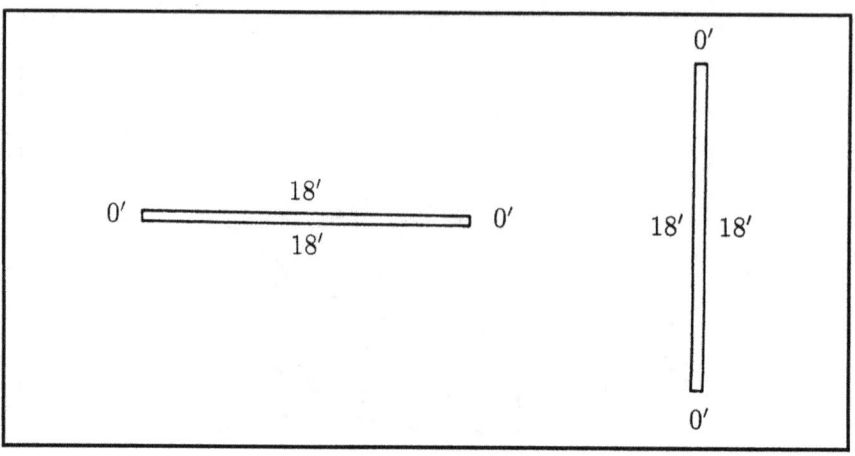

Figure 8.2. The extreme cases of long, narrow pens certainly do not maximize area.

8.4. The Double Rectangular Pen Problem

Find the biggest adjacent pair of identical rectangular pens you can make with 36 feet of fencing. See Figure 8.3.

Solution. I'll bet you can't guess the solution this time. The quantity to be maximized is the area A. For the variable we use the length ℓ of each pen. If w denotes the width of each pen, then the fixed perimeter is given by $3\ell + 4w = 36$. Therefore,

$$w = \frac{36 - 3\ell}{4} = 9 - \frac{3}{4}\ell, \qquad \text{and}$$

$$A = 2\ell w = 2\ell(9 - \frac{3}{4}\ell) = 18\ell - \frac{3}{2}\ell^2.$$

Again the extreme cases are obviously no good, and we turn to the critical cases:

$$0 = A' = 18 - 3\ell = 3(6 - \ell).$$

Thus

$$\ell = 6, \quad w = 9 - \frac{3}{4}\ell = 4\frac{1}{2}.$$

Each pen should be exactly $6 \times 4\frac{1}{2}$, exactly this much longer than wide to take full advantage of the wall they share.

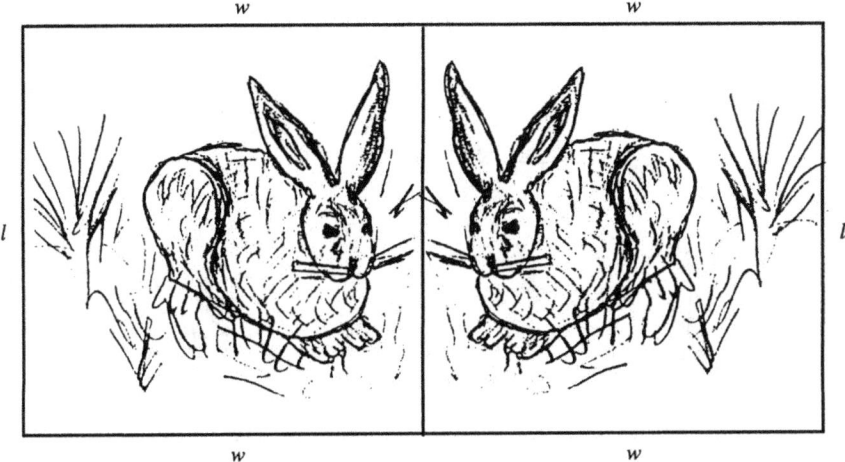

Figure 8.3. Find the biggest adjacent pair of identical rectangular pens you can make with 36 feet of fencing. (Illustration by the author's mother, Brennie Morgan.)

8.5. The Circular Pen Problem

Find the biggest pen (not necessarily rectangular) you can make with 36 feet of fencing.

Solution. Since the infinitude of possible shapes cannot be described by a single variable x, this problem exceeds the capabilities of our current methods. It properly belongs to the infinite-dimensional calculus, or the *calculus of variations*. As you might guess, the answer is the circle of Figure 8.4. Since its perimeter is 36, its radius r is $36/2\pi \approx 5.73$, and its area $\pi r^2 \approx 103$, bigger than the comparable square of area 81.

8.6. The Round Sphere Problem

What shape in space most efficiently encloses volume?

Solution. Although well beyond our ability to prove it, the round sphere is the ideal shape, as discovered by bubbles long before humanity inhabited the earth, and proved by Schwarz in 1884. See Figure 8.5.

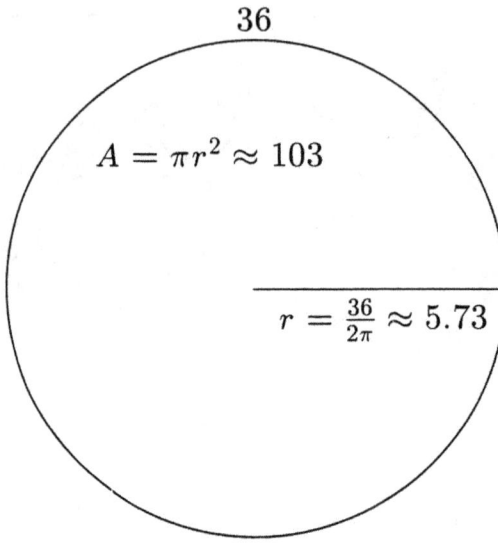

36

$$A = \pi r^2 \approx 103$$

$$r = \frac{36}{2\pi} \approx 5.73$$

Figure 8.4. The circle encloses the largest area of a given perimeter.

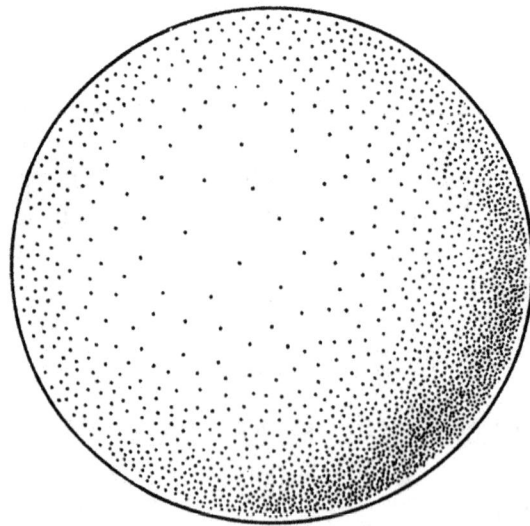

Figure 8.5. The round soap bubble provides the most efficient way to enclose a given volume of air.

8.7. The Double Bubble Problem

What shape in space most efficiently encloses two separate volumes?

The ideal shape is a double soap bubble as in Figure 8.6, which forms when two soap bubbles come together. This result was finally proved by a team including the author in 2000. We had to rule out some other crazy possibilities, as suggested by the strange computer-generated double bubble in Figure 8.7, in which one bubble wraps around the other like an innertube. Also in 2000, a group of undergraduate students led by Ben Reichardt extended the proof to four-dimensional double bubbles in four-dimensional space. (For more information, see my article in the November 2000 Math Horizons or my Math Chat column archive at MathChat.org.)

All of these results can be traced to the proof in 1990 of the planar double bubble theorem, which says that the analogous double pen of Figure 8.8 is the most efficient way to fence in two areas. This seminal result too was proved and published by a group of undergraduates: Joel Foisy, Manuel Alfaro, Jeffrey Brock, Nickelous

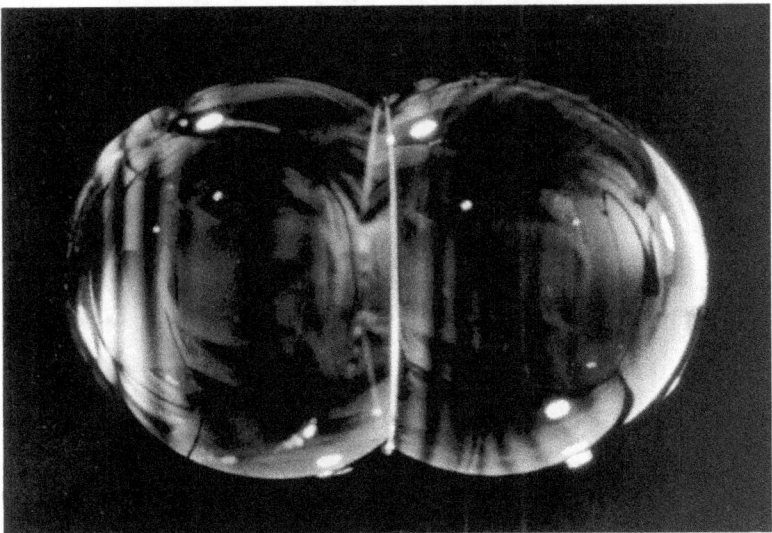

Figure 8.6. It was finally proved in 2000 that the familiar double soap bubble is the most efficient way to enclose two separate volumes of air. (Photo by Jeremy Ackerman, Washington University '96.)

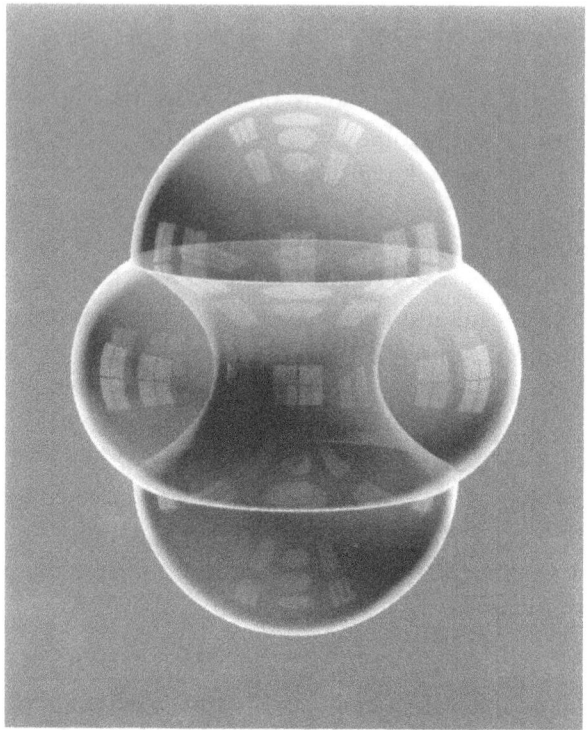

Figure 8.7. A new computer-generated double bubble, which however has more surface area than the standard double bubble of Figure 8.6. (John Sullivan, University of Minnesota.)

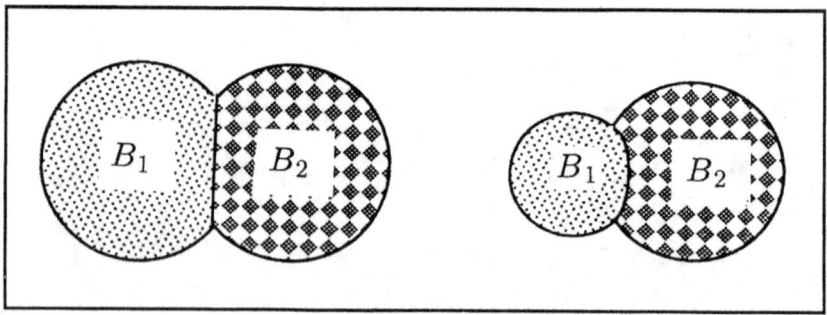

Figure 8.8. This double pen type is the most efficient way to fence in two areas. This fact was proved just in 1990 by a group of Williams undergraduates: Manuel Alfaro, Jeffrey Brock, Joel Foisy, Nickelous Hodges, and Jason Zimba.

Hodges, and Jason Zimba. The work of these students manifests a national resurgence of activity, with undergraduates proving theorems, writing up results for publication, and speaking at mathematics meetings.

8.8. The Cylindrical Can Problem

What are the dimensions of a 1000 cm^3 cylindrical can which uses the least amount of material?

Solution. The quantity to be minimized is the surface area A. For the variable, we could use the radius r or the height h; then the unused variable is determined by the prescribed volume $V = \pi r^2 h = 1000$. (See Figure 8.9.) Let's use r as the variable. Then $h = 1000/\pi r^2$, and

$$
\begin{aligned}
A &= f(r) = A_{\text{top}} + A_{\text{bottom}} + A_{\text{sides}} \\
&= \pi r^2 + \pi r^2 + 2\pi r h \\
&= 2\pi r^2 + \frac{2000}{r}.
\end{aligned}
$$

The extreme cases occur as $r \to 0$ or $r \to \infty$ (see Figure 8.10).

Perhaps your physical intuition tells you that the extreme cases, tall skinny cans and short fat cans, use lots of material. Our formula for $A = f(r)$ also tells us that as $r \to 0$ or $r \to \infty$, $A \to \infty$. Therefore the minimum will occur in between where $f' = 0$,

$$
f' = 4\pi r - \frac{2000}{r^2} = 0,
$$

which implies that

$$
r^3 = \frac{1000}{2\pi}, \qquad r = \frac{10}{\sqrt[3]{2\pi}}.
$$

Since f and f' are always well defined, this is the only possibility, and the best dimensions are

$$
r = \frac{10}{\sqrt[3]{2\pi}}, \qquad h = \frac{1000}{\pi r^2} = \frac{1000}{\pi r^3} r = \frac{1000}{500} r = 2r = \frac{20}{\sqrt[3]{2\pi}},
$$

with the height equal to the diameter. *Are* soup cans this shape?

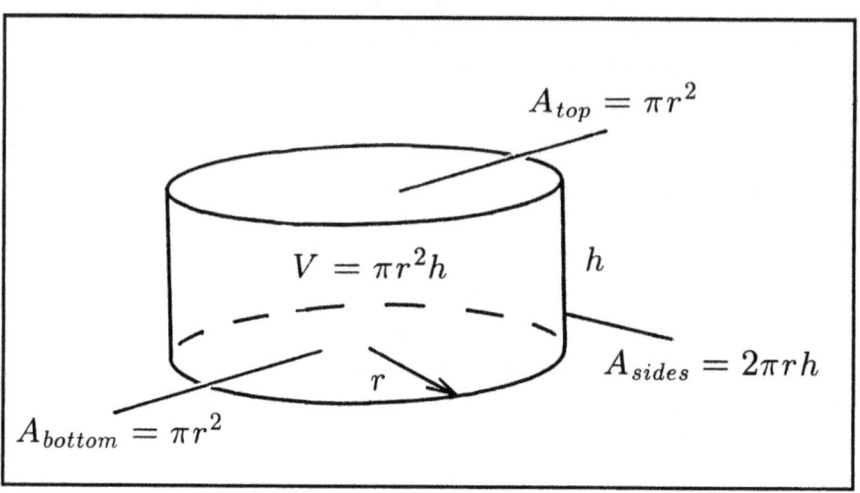

Figure 8.9. A cylindrical can.

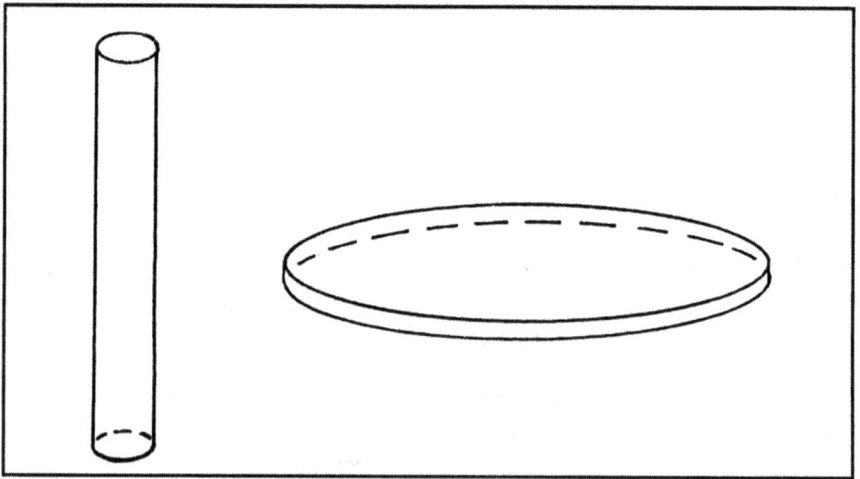

Figure 8.10. The extreme cases occur as $r \to 0$ (tall skinny can) or $r \to \infty$ (short fat can).

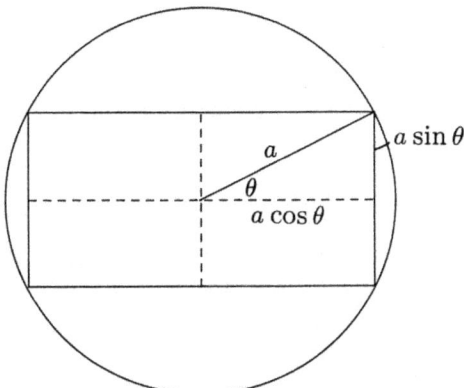

Figure 8.11. Rectangles inscribed in a circle of radius a are best described by an angle θ.

8.9. The Inscribed Rectangle Problem (Using Angles)

Find the largest rectangle inscribed in a given circle.

Solution. The quantity to be maximized is the area A. For the variable, we could use the length ℓ or the width w, but whenever possible it is easier to use an angle θ (see Figure 8.11).

Of course the maximum area will depend on the size of the circle: we assume that its radius a is given. Since a is the hypotenuse of a right triangle, the side opposite the angle θ is $a \sin \theta$ and the adjacent side is $a \cos \theta$ (see Figure 8.12).

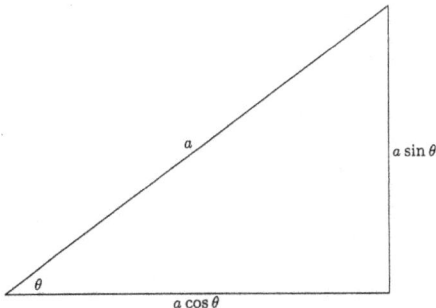

Figure 8.12. If a is the hypotenuse of a right triangle, the side opposite the angle θ is $a \sin \theta$ and the adjacent side is $a \cos \theta$.

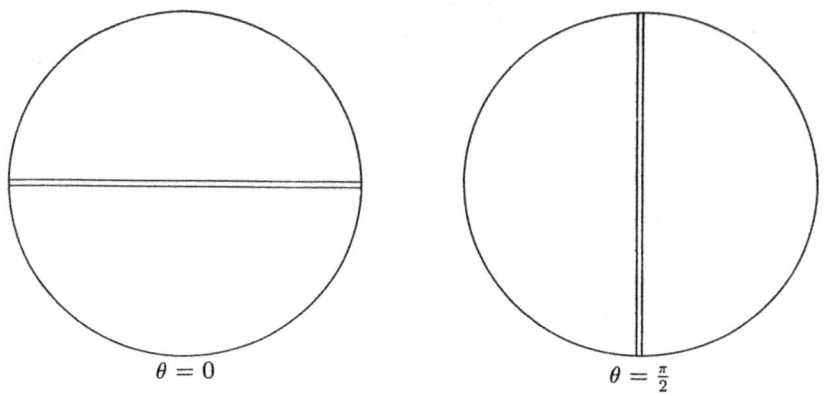

Figure 8.13. The stupid cases $\theta = 0$ and $\theta = \pi/2$.

Hence the length ℓ of the rectangle is $2a \cos \theta$, the width w is $2a \sin \theta$, and the area

$$A = \ell w = (2a \cos \theta)(2a \sin \theta) = 4a^2 \cos \theta \sin \theta.$$

The extreme cases $\theta = 0$ and $\theta = \pi/2$ give area 0, clearly not the maximum! See Figure 8.13. When the derivative $A' = 0$, by the Product Rule

$$\begin{aligned}
0 = A' &= 4a^2(-\sin \theta) \sin \theta + 4a^2 \cos \theta \cos \theta \\
&= 4a^2(-\sin^2\theta + \cos^2\theta), \\
\sin^2\theta &= \cos^2\theta, \\
\sin \theta &= \pm \cos \theta.
\end{aligned}$$

Since $0 \le \theta \le \pi/2$, both $\sin \theta$ and $\cos \theta$ are positive and hence

$$\begin{aligned}
\sin \theta &= \cos \theta \\
\theta &= \frac{\pi}{4}.
\end{aligned}$$

Since A and A' are always well defined, this must be the maximum:

$$\begin{aligned}
\ell &= 2a \cos \frac{\pi}{4} = \frac{2a}{\sqrt{2}} = a\sqrt{2}, \\
w &= 2a \sin \frac{\pi}{4} = \frac{2a}{\sqrt{2}} = a\sqrt{2}.
\end{aligned}$$

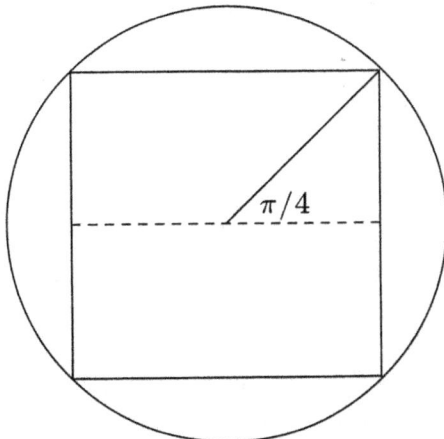

Figure 8.14. The largest rectangle inscribed in a circle is a square.

The solution is a square as you might have guessed. See Figure 8.14.

Remark. An alternative method just notices that

$$A = 4a^2 \cos\theta \sin\theta = 2a^2 \sin 2\theta,$$

which we know has its maximum value of $2a^2 \cdot 1$ when $2\theta = \pi/2$, when $\theta = \pi/4$, as before.

8.10. Advice for Working Problems

1. Know the three steps from the beginning of this section.

2. Be wise in selecting the variable to adjust. Use r for cylinders, an angle θ whenever convenient (see Problems 8.8 and 8.9).

3. Know the formulas, such as

   - the area of a triangle: $A = \frac{1}{2}bh$,
   - the area of a circle: $A = \pi r^2$,
   - the circumference of a circle: $2\pi r$,

   and the formulas for a cylinder and for a right triangle in Figures 8.9 and 8.12.

Exercises 8

1. What is the shape of an open-top cylindrical can of least material for given volume V?

2. What is the shape of the cheapest open-top cylindrical can of given volume V if the base material costs half as much as the side material?

3. A student with 12 feet of fence wants to enclose on three sides the largest possible rectangular rabbit pen against one outside wall of her dormitory. What should its dimensions be?

4. Another student has two straight 4-foot pieces of fence for a triangular pen along the dormitory. How much area can she enclose?

5. Find the shape of the rectangle of greatest area that can be enclosed in a semicircle. Assume that the sides are horizontal and vertical.

6. A farmer pays a boy $3 per hour to sell peas after school from 3 p.m. on. (a) He sells $12t - \frac{3}{2}t^2$ pounds, where t is the number of hours since 3 p.m. The farmer makes $1/pound. How long should the farmer have the boy work? (b) Suppose he sells $2t - \frac{1}{4}t^2$ pounds?

7. A farmer has 1000 feet of fencing to use to enclose a rectangular pen. (a) Find the dimensions of the pen that yield a maximum area enclosed; (b) find the dimensions that give maximum area if the farmer uses the side of a large barn as one side of the pen.

8. Find the point on the parabola $y = x^2/2$ that is closest to the point (4,1).

9. A long rectangular sheet of metal 8 feet wide is to be made into a triangular water trough by folding it in half along the length. How wide across the top should it be so that it holds the most water?

10. If you pick your oranges on January 1 (when each tree yields 130 pounds of oranges) you can get 16 cents per pound at market. The yield increases at a rate of 5 pounds per tree per week, but the market price decreases 1/2 cent per pound per week. When should you pick the oranges to make the most money?

11. A rectangular box with two square opposite ends is to hold 8000 cubic inches.

(a) Find the dimensions of the box so that the least amount of material is used;

(b) Find the dimensions of the cheapest box if the rectangular sides cost 15 times as much per square inch as the top, bottom, and square ends.

12. A rectangular cardboard poster for advertising calculus in Berlin is to contain 216 square inches of printed material with 2 inch margins at the sides and 3 inch margins at the top and bottom. Find the dimensions of the poster to use the least possible amount of cardboard.

13. An advertiser learns that she can attract 200 new customers every time a commercial is run. Unfortunately, it also seems that she loses $5n^2$ customers for running the ad n times, as people become annoyed at having their program interrupted. How many times should she run the ad to get the most customers?

14. If a man throws a ball at an angle θ, the distance it will travel is equal to $(v \cos \theta)(2\frac{v}{g} \sin \theta)$, where v is the speed he throws it and g is the acceleration of gravity. For given v and g, what angle should he throw it to maximize the distance?

15. (from a Russian Olympiad via Sergei Chmutov) What is the maximum overlap area of two right isosceles triangles with legs of length 1 on the x-axis, facing in opposite directions:

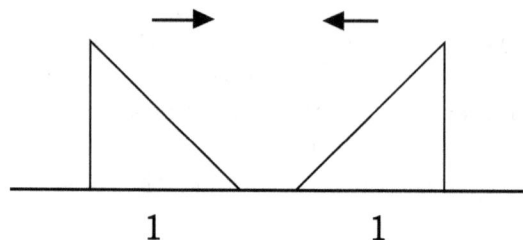

9

Exponentials and Logarithms

$\log_a x$ ("log base a of x") is the power you have to raise a to in order to get x. For example, $\log_2 8 = 3$ because $2^3 = 8$. $\log_{10} 2 = .3010\ldots$ because $10^{.3010\ldots} = 2$. $\log_4 1/8 = -3/2$, because

$$4^{-\frac{3}{2}} = \frac{1}{(\sqrt{4})^3} = \frac{1}{2^3} = \frac{1}{8}.$$

In other words, $a^{\log_a x} = x$.

Examples.

$$\log_9 27 \quad = \tfrac{3}{2} \qquad \log_{\frac{1}{2}} 4 \quad = -2$$
$$2^{\log_2 100} \quad = 100 \qquad 2^{\log_3 100} \quad \text{cannot be simplified.}$$

The basic truism that $a^{\log_a x} = x$ is more a matter of logic than mathematics. The mother of "the man whose mother is Barbara" is …Barbara! Likewise, a raised to "the power you raise a to to get x" is …x! That is, $a^{\log_a x} = x$.

Similarly, the man whose mother is the mother of David is …David! Likewise, the power you raise a to to get a^x is …x! That is, $\log_a a^x = x$. In particular, taking $a = e$, $e^{\ln x} = x$ and $\ln e^x = x$.

If one assumes, as we will, that both a and x are positive and $a \neq 1$, then $\log_a x$ is a well determined real number. If $a > 1$,

$$\log_a x \text{ is } \begin{cases} \text{positive} & \text{if } x > 1 \\ 0 & \text{if } x = 1 \quad (\text{since } a^0 = 1) \\ \text{negative} & \text{if } x < 1; \end{cases}$$

on the other hand if $a < 1$,

$$\log_a x \text{ is } \begin{cases} \text{negative} & \text{if } x > 1 \\ 0 & \text{if } x = 1 \\ \text{positive} & \text{if } x < 1. \end{cases}$$

9.1. Logarithm Rules

Logarithms obey the following rules:

Product Rule:	$\log_a xy = \log_a x + \log_a y$
Quotient Rules:	$\log_a \frac{x}{y} = \log_a x - \log_a y, \quad \log_a \frac{1}{x} = -\log_a x$
Power Rule:	$\log_a x^p = p \cdot \log_a x$
Base Change Rule:	$\log_b x = \frac{\log_a x}{\log_a b}$

Note that there is no way to simplify $\log_a(x + y)$ since

$$\log_a(x + y) \neq \log_a x + \log_a y.$$

Examples 9.1. By the Product and Power rules,

$$\log_2 3^5 7^9 = \log_2 3^5 + \log_2 7^9 = 5\log_2 3 + 9\log_2 7;$$
$$2^{\log_4 100} = 4^{\frac{1}{2}\log_4 100} = 4^{\log_4 100^{\frac{1}{2}}} = 100^{\frac{1}{2}} = 10;$$

by the Base Change Rule,

$$\log_4 100 = \frac{\log_2 100}{\log_2 4} = \frac{\log_2 10^2}{2} = \frac{2\log_2 10}{2} = \log_2 10.$$

The most frequently used bases are 10 ("common logarithms," sometimes written just $\log x$) and another number $e = 2.71828\ldots$

("natural logarithms," written $\ln x$). It is time to introduce this constant e of Euler's, for it plays a very important simplifying role in calculus.

9.2. The Definition of e

Before defining e, I'd like to emphasize how important it will be in describing exponential growth, as for money in the bank with continuous interest compounding or for bacteria populations under unrestricted growth.

To define e, suppose you start with the simplest amount of money in the bank (\$1) for the simplest amount of time (1 year) at the mathematically simplest though unrealistic interest rate (100%), compounded continuously. How much money do you end up with by the end of the year? The answer is: e dollars! In fact, we'll make that our definition of e.

To get a formula for e, divide the year into n periods, at the end of each period pay interest of $100\%/n$ (*i.e.* multiply the bank balance by $1+1/n$), do this for n periods to produce a final balance of $\$1 \cdot (1+1/n)^n$, and take the limit as $n \to \infty$ ("continuous compounding"):

$$e = \lim_{n \to \infty} \left(1 + \frac{1}{n}\right)^n = 2.71828\dots$$

For example, $e \approx (1 + 1/1000)^{1000}$. A still better approximation would be $e \approx (1 + 1/1000000)^{1000000}$. So in one year the money gets multiplied by e. In t years, it gets multiplied by e^t. At a general interest rate k, since *e.g.* doubling the interest rate is the same as doubling the time, it gets multiplied by e^{kt}. Finally, if you start with y_0 ("y sub zero" or more usually the shorter "y naught") dollars, you end with $y_0 e^{kt}$. We have the following law.

9.3. Law of Exponential Growth

A quantity y growing continuously at a rate proportional to itself $(dy/dt = ky)$ is given by the equation

$$y = y_0 e^{kt},$$

where y_0 is the starting value of y at time $t = 0$ and k is the growth rate.

We will see in Chapter 10 that money invested at interest rate k, compounded continuously, is just one of many important examples of exponential growth (so called because the variable t occurs as an exponent). Since the rate of change dy/dt caused by the interest is proportional to the amount y, i.e., $y' = ky$,

$$(y_0 e^{kt})' = k(y_0 e^{kt}).$$

In particular, if $y_0 = 1$ and $k = 1$,

$$(e^t)' = e^t.$$

We add this to our list of important rules, returning from using t to using x as the variable.

9.4. Exponential Rule

$$(e^x)' = e^x$$
$$(e^u)' = e^u u'.$$

The function e^x is its own derivative, i.e., its value e^x and its rate of change are always equal. At each point of the graph $y = e^x$, the height and the slope are equal.

Using the Chain Rule, we can now compute

$$\left(e^{10x^2}\right)' = 20 x \cdot e^{10x^2},$$
$$\left(e^{\sqrt{x}}\right)' = \frac{1}{2\sqrt{x}} e^{\sqrt{x}}.$$

9.5. Exponential Rule (base a)

$$(a^x)' = (\ln a) a^x ;$$
$$(a^u)' = (\ln a) a^u u' .$$

Examples 9.2.

$$(2^x)' = (\ln 2) \cdot 2^x,$$
$$(2^{x^2})' = (\ln 2)(2x) \cdot 2^{x^2},$$
$$(e^x)' = (\ln e) \cdot e^x = e^x.$$

Notice how much easier the formula is for the natural base $a = e$.

Proof. Since

$$a^x = \left(e^{\ln a}\right)^x = e^{(\ln a)\,x},$$

$$(a^x)' = \left(e^{(\ln a)\,x}\right)' = (\ln a) \cdot e^{(\ln a)\,x} = (\ln a) \cdot a^x.$$

Notice how we treat $\ln a$ as any other constant.

Recall that since $(x^n)' = nx^{n-1}$ ($n \neq 0$), but $(x^0)' = (1)' = 0$, x^{-1} never turned up as the derivative of a power of x. The derivative of the natural logarithm $\ln x$ turns out to be the missing x^{-1}.

9.6. Logarithm Rule

$$(\ln x)' = \tfrac{1}{x}; \qquad\qquad (\ln u)' = \tfrac{1}{u}\,u';$$

$$(\log_a x)' = \tfrac{1}{\ln a}\tfrac{1}{x}; \qquad (\log_a u)' = \tfrac{1}{\ln a}\tfrac{1}{u}\,u'.$$

Example 9.3. If $y = \ln(7x^2 + 4)$, then

$$y' = \frac{1}{7x^2 + 4}(14x) = \frac{14x}{7x^2 + 4}.$$

Proof. Suppose $y = \ln x$. Then $x = e^y$. Implicit differentiation yields $1 = e^y dy/dx$. Hence $dy/dx = e^{-y} = 1/e^y = 1/x$. To differentiate $\log_a x$, use the formula

$$\log_a x = \frac{\ln x}{\ln a}.$$

Hence

$$(\log_a x)' = \frac{1}{\ln a}\frac{1}{x}.$$

Exercises 9

Compute.

1. $\log_2 16$

2. $\log_3 81$

3. $\log_8 32$

4. $\log_{25} 125$

5. $\log_4 \frac{1}{16}$

6. $\log_{27} \frac{1}{3}$

7. $\log_{1/2} 4$

8. $\log_{2/3} \frac{9}{4}$

9. $\log_{2/5} \frac{4}{25}$

10. $\log_{.1} .001$

Simplify or label "cannot be simplified."

11. $3^{\log_3 4}$

12. $5^{\log_5 32}$

13. $3^{\log_9 5}$

14. $25^{\log_5 9}$

15. $3^{\log_4 3}$

16. $2^{\log_6 40}$

17. $\log_2(x + y)$

18. $(\log_3 x)(\log_3 y)$

19. $(\log_3 x)(\log_5 y)$

20. $\log_3 x + \log_3 y$

21. $\log_3 x + \log_5 y$

22. $(\log_3 5)(\log_5 3)$

Differentiate.

23. e^{5x}

24. e^{7x^2+x}

25. $e^{\sqrt[3]{x}}$

26. $e^{x^{-\frac{5}{3}}}$

27. $e^{\sin x}$

28. $e^{\cos x}$

29. $e^x \sin x$

30. $e^{x^3} \cos x^2$

31. $e^{\sin(x^2-5x-6)}$

32. e^{e^x} (This means $e^{(e^x)}$.)

33. $(e^e)^x$

34. x^e

35. 7^x

36. $\left(\dfrac{1}{7}\right)^x$

37. 2^{x^2+6x-5}

38. $3^{\sqrt{x}}$

39. 5^{x^3}

40. $\sqrt{5}^{x^2}$

41. $e^{\frac{x+1}{x-1}}$

42. $\ln(x^2 + 1)$

43. $\ln(x^4 + 1)$

44. $\ln(x^4 - x^2 + 1)$

45. $\ln(e^x + 1)$

46. $\ln e^x$

47. $\ln(\cos^2 x + 1)$

48. $\cos \ln x$

49. $\ln^2 x$ (This means $(\ln x)^2$.)

50. $\ln \dfrac{x^2 + 1}{x^2 + 2}$

51. $\ln 2x$

52. $\ln 3x$

53. $\log_{10}(x^2 + 7)$

54. $\ln \ln x$

55. $\ln \ln \ln x$

56. $\sqrt{\ln x}$

10

Exponential Growth and Decay

Since the Law of Exponential Growth 9.3 $y = y_0 e^{kt}$ holds for any quantity y growing at a rate proportional to itself ($dy/dt = ky$), it models many physical phenomena such as population growth, money in the bank, and radioactive decay. We will sometimes use Newton's notation with a dot $\dot{y}$ for time derivatives:

$$\dot{y} = dy/dt.$$

In this notation, if $\dot{y} = ky$, then $y = y_0 e^{kt}$.

10.1. Population

In the simplest model, population y grows at a rate proportional to itself, $\dot{y} = ky$: the number of new births (minus the number of deaths) is some fraction of the current population; k is the instantaneous growth rate.

Example 10.1. Suppose the US population grows continuously at the instantaneous rate of 3% per year. If the population is currently 275 million, what will it be in one year? In ten years?

Solution. Since $\frac{dy}{dt} = .03y$, $y = y_0 e^{kt} = 275 e^{.03t}$ million. In one year, $y(1) = 275 e^{.03} \approx 283$ million. In ten years, $y(10) = 275 e^{.3} \approx 371$ million. These are increases of about 3.05% and 35%, which are more than $1 \times 3\%$ and $10 \times 3\% = 30\%$, because the increases continuously build on each other.

10.2. Money in the Bank

Since money y in the bank grows at a rate proportional to itself ($\dot{y} = ky$, where k is the annual interest rate, compounded continuously), y must equal $y_0 e^{kt}$. This occurs as a limit of frequent compounding:

compounding	yields
annually for t years:	$y = y_0(1 + k)^t$
monthly for t years:	$y = y_0\left(1 + \frac{k}{12}\right)^{12t}$
daily for t years:	$y = y_0\left(1 + \frac{k}{365}\right)^{365t}$
kn periods for t years:	$y = y_0\left(1 + \frac{k}{kn}\right)^{knt} = y_0\left(\left(1 + \frac{1}{n}\right)^n\right)^{kt}$
continuously for t years:	$y = y_0 e^{kt}$, since $e = \lim\limits_{n \to \infty}\left(1 + \frac{1}{n}\right)^n$

Example 10.2. How long will it take money in the bank to double at 10% interest compounded continuously? Compounded annually?

Solution. For interest compounded continuously, $y = y_0 e^{kt} = y_0 e^{.1t} = 2y_0$, $e^{.1t} = 2$, $.1t = \ln 2$, $t = 10 \cdot \ln 2 \approx 7$ years. This exhibits bankers' "Rule of 70," which says that the doubling time can be estimated by dividing the interest rate into 70.

For interest compounded annually, $y = y_0(1 + k)^t = y_0(1.1)^t$. At doubling, $(1.1)^t = 2$, $t \ln 1.1 = \ln 2$, $t = \ln 2 / \ln 1.1 \approx 7.3$. The balance would not double until the eighth interest payment.

10.3. Radioactive Decay and Half-life

The amount y of a radioactive material present *decays* at a rate proportional to the amount present. The *half-life* h is the time in which half of the material decays. In this case the law of exponential growth for the amount y of radioactive material remaining at time t

simplifies to

$$y = y_0 \left(\frac{1}{2}\right)^{\frac{t}{h}}. \qquad\qquad (10.1)$$

After t/h half-lives, y_0 has been halved t/h times.

Example 10.3. If uranium-235 has a half-life of .71 billion years, what fraction will decay in one billion years?

Solution. The amount of material remaining after one billion years is given by

$$y = y_0 \left(\frac{1}{2}\right)^{\frac{1\ \text{billion}}{.71\ \text{billion}}}.$$

The fraction remaining after one billion years is

$$\frac{y}{y_0} = \left(\frac{1}{2}\right)^{\frac{1}{.71}} \approx 38\%.$$

The fraction decayed is

$$1 - \left(\frac{1}{2}\right)^{\frac{1}{.71}} \approx 62\%.$$

Derivation of Formula (10.1)

The general exponential growth law would say that $y = y_0 e^{kt}$. After one half-life, half remains:

$$\frac{1}{2} y_0 = y_0 e^{kh},$$

so that

$$\frac{1}{2} = e^{kh}, \quad e^k = \left(\frac{1}{2}\right)^{\frac{1}{h}}.$$

Therefore $y = y_0 e^{kt} = y_0 (\frac{1}{2})^{(t/h)}$.

10.4. Summary Table for Quantity y Growing at Rate Proportional to Itself ($\dot{y} = ky$)

Population	$y = y_0 e^{kt}$	y_0 initial population k growth rate
Money	$y = y_0 e^{kt}$	y_0 initial amount k cont. interest rate
Radioactive material	$y = y_0 \left(\frac{1}{2}\right)^{\frac{t}{h}}$	y_0 initial amount h half-life

Notes.

1. For radioactive decay, after t/h half-lives, y_0 has been multiplied by $1/2$, t/h times.

2. The amount is given by y, the fraction remaining by y/y_0, and the fraction decayed by $1 - (y/y_0)$.

3. To solve for t, take ln of both sides.

10.5. Rates of Growth (Continuation of 7.3)

Exponential growth e^{ax} ($a > 0$) overwhelms any power x^p of x as x gets large. (Populations can grow very fast.) In the terminology of Section 7.3, x^p grows slower than e^{ax}:

$$x^p \ll e^{ax},$$

which just means that

$$\lim_{x \to +\infty} \frac{x^p}{e^{ax}} = 0.$$

Proof sketch. We will show that if

$$y = \frac{e^{ax}}{x^p},$$

then $\lim_{x \to +\infty} y = +\infty$.

By the Product Rule,

$$y' = (e^{ax}x^{-p})' = ae^{ax}x^{-p} + e^{ax}(-px^{-p-1})$$
$$= e^{ax}x^{-p}(a - \frac{p}{x}) = (a - \frac{p}{x})y \geq \frac{a}{2}y,$$

once $x \geq 2p/a$. We know that if $y' = (a/2)y$, then $y = ce^{(a/2)x}$ and $\lim_{x \to +\infty} y = +\infty$; since $y' \geq (a/2)y$, y grows even faster and certainly $\lim_{x \to +\infty} y = +\infty$, as desired.

Similarly $\ln x$ is overwhelmed by any power of x (even $x^{1/100}$). In summary,

$$\ln x \ll x^{\frac{1}{100}} \ll x \ll x^{100} \ll e^{\frac{x}{100}} \ll e^x \ll e^{100x}.$$

Exercises 10

1. Suppose the population of China grows continuously at an instantaneous rate of 5% per year. If the population is currently 1.5 billion, what will it be in 100 years?

2. A Pseudomonas bacteria culture grows continuously at an instantaneous rate of 4% a minute. If the population is currently 1000, what will it be in one hour? Ten hours?

3. How long will it take money in the bank to grow from $1000 to $1,000,000 at 10% interest (a) compounded continuously? (b) compounded annually?

4. How long will it take money in the bank to triple at 7% interest compounded continuously? At 14%?

5. Uranium-238 has a half-life of 4.5 billion years. What fraction will decay in one billion years?

6. Two hundred thirty-eight grams of uranium-238 (half-life 4.5 billion years) contain Avogadro's number, about 6×10^{23}, atoms. How many atoms will decay in the first year? In the first second?

7. Carbon-14 has a half-life of 5730 years. What fraction will decay in 1000 years? How long does it take for 10% to decay? 20%? 90%?

8. The amount of carbon-14 (half-life 5730 years) left after 10,000 years is 5 grams. How much was there to start with?

9. Last year the spider population in my office was growing at the alarming instantaneous rate of 50% (time measured in days). If there are 100 spiders to start, how many days will it take for my office to be crawling with a million spiders?

10. In the first hour a culture of *E. coli* bacteria grows from 1000 to 7500. Predict the population after one day.

11. A Petri dish culture is begun with a single *S. aureus* bacterium. Each bacterium divides every 20 minutes. How long will it take to obtain the desired culture of one trillion (10^{12}) bacteria?

12. Arrange the following functions in a chain of $\ll$.

$$e^{2x}, \ln x, 100x^4, \frac{1}{100}e^x, \frac{1}{100}e^{3x}, \frac{1}{100}\sqrt{x}$$

Compute the following limits.

13. $\lim\limits_{x\to+\infty} \dfrac{e^x}{e^{2x}+1}$

14. $\lim\limits_{x\to+\infty} \dfrac{3x^2+2x+1}{4x^2+10x+7}$

15. $\lim\limits_{x\to+\infty} \dfrac{5x+2e^x+7e^{2x}}{\ln x + e^{2x}}$

16. $\lim\limits_{x\to+\infty} \dfrac{5x+2e^x}{\ln x + e^{2x}}$

17. $\lim\limits_{x\to+\infty} \dfrac{\sqrt{e^x+\ln x}}{e^{\frac{x}{2}}}$

18. $\lim\limits_{x\to+\infty} \dfrac{e^x}{x^{10^{20}}}$

19. a. $\displaystyle\lim_{x\to0^+} x - 4x^{-1}$ b. $\displaystyle\lim_{x\to0^-} x - 4x^{-1}$

20. a. $\displaystyle\lim_{x\to4^+} \frac{x^2 - 4}{x - 4}$ b. $\displaystyle\lim_{x\to4^-} \frac{x^2 - 4}{x - 4}$

21. $\displaystyle\lim_{x\to+\infty} \frac{\sqrt{x}}{x^2 + 1}$

11

The Second Derivative
and Curve Sketching

If $y = f(t)$ denotes mileage, the derivative $\dot{y} = dy/dt = f'(t)$ denotes velocity, and the derivative of the derivative or *second derivative*

$$\ddot{y} = \frac{d^2y}{dt^2} = f''(t)$$

denotes acceleration, the rate of change of the velocity.

Just as the sign of the derivative tells whether the graph $y = f(t)$ slopes up or down, the sign of the second derivative tells whether the graph is concave upward or downward as in Figure 11.1, because it tells whether the slope is increasing or decreasing. See Figures 11.1 and 11.2.

11.1. Curve Sketching

In sketching a graph of $y = f(x)$, the following information facilitates a quicker and better picture than you could get by just plotting points.

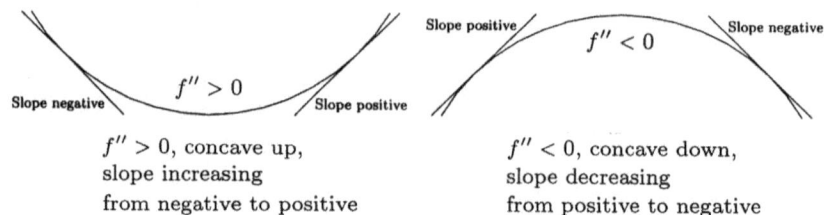

$f'' > 0$, concave up,
slope increasing
from negative to positive

$f'' < 0$, concave down,
slope decreasing
from positive to negative

Figure 11.1. The sign of the second derivative tells whether the graph is concave upward or downward, because it tells whether the slope is increasing or decreasing.

(a) What happens as x goes to the endpoints or to $\pm\infty$ or to places where a denominator goes to 0?

(b) Where is the function 0? Positive? Negative?

(c) Where is the derivative 0? Positive? Negative? (This provides information on the slope and extrema.)

(d) Where is the second derivative 0? Positive? Negative? (This provides information on the concavity.)

Here is some new terminology that will come up in this chapter:

Relative maximum. If no *nearby* values of f are larger, f has a relative maximum at x. (Guaranteed if $f'(x) = 0$ and $f''(x) < 0$.)

Relative minimum. If no *nearby* values of f are smaller, f has a relative minimum at x. (Guaranteed if $f'(x) = 0$ and $f''(x) > 0$.)

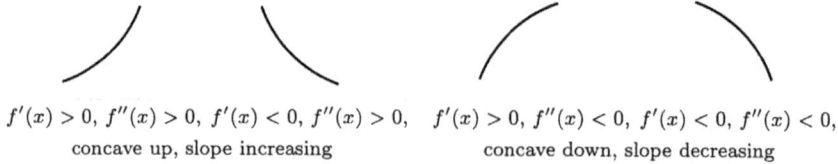

$f'(x) > 0$, $f''(x) > 0$, $f'(x) < 0$, $f''(x) > 0$,
concave up, slope increasing

$f'(x) > 0$, $f''(x) < 0$, $f'(x) < 0$, $f''(x) < 0$,
concave down, slope decreasing

Figure 11.2. The four possible shapes of a piece of graph. If $f'(x) > 0$, the graph slopes upward. If $f''(x) > 0$, the graph is concave up.

To distinguish them from relative maxima and minima, genuine maxima and minima are often called *absolute maxima and minima.*

Inflection point. An inflection point is a point where f'' changes sign, i.e., concavity changes from up to down or from down to up. (Often occurs if $f''(x) = 0$.)

Asymptote. An asymptote is a line approached by the graph.

Figures 11.3(a–f) illustrate important features of graphs. Figure 11.3(a) shows $y = 2x + 1$, which you immediately recognize from the formula as a line $y = mx + b$ with slope $m = 2$ and y-intercept $b = 1$ (*i.e.*, it crosses the y-axis at $y = 1$).

Figure 11.3(b) shows $y = -x^2 + 2x + 1$, which you immediately recognize from the formula as a parabola, since only x^2's, x's, and constants occur. Since the x^2 term grows fastest, the negative sign on the x^2 means that as $x \to \pm\infty$, $y \to -\infty$. Hence it must be a downward parabola. Another way to see this is that the second derivative $y'' = -2$, so that the graph is concave downward. The most important feature is the maximum, which occurs where $0 = y' = -2x + 2$, so that $x = 1$, and hence by the formula $y = 2$.

Figure 11.3(c) shows the graph of a cubic (with an x^3 term). Since the x^3 term grows fastest, as $x \to +\infty$, $y \to +\infty$, and as $x \to -\infty$, $y \to -\infty$. The relative extrema can be found by setting $y' = 0$. The concavity $y'' = 6x - 12 = 6(x - 2)$ changes from negative to positive at $x = 2$, $y = 2$, the inflection point. Just as a parabola wiggles once, with one relative extremum and no inflection points, a cubic wiggles twice, with up to two relative extrema and exactly one inflection point.

Figure 11.3(d) shows the graph of a quartic (with an x^4 term). Since the x^4 term grows fastest, as $x \to \pm\infty$, $y \to +\infty$. The relative extrema and inflection points occur where y' and y'' are zero. A quartic wiggles up to three times, with up to three relative extrema and up to two inflection points.

Figure 11.3(e) graphs a function with a fraction, but because the denominator is never 0, the function never blows up. As $x \to \pm\infty$, $y \to 1$, and the line $y = 1$ is a horizontal asymptote.

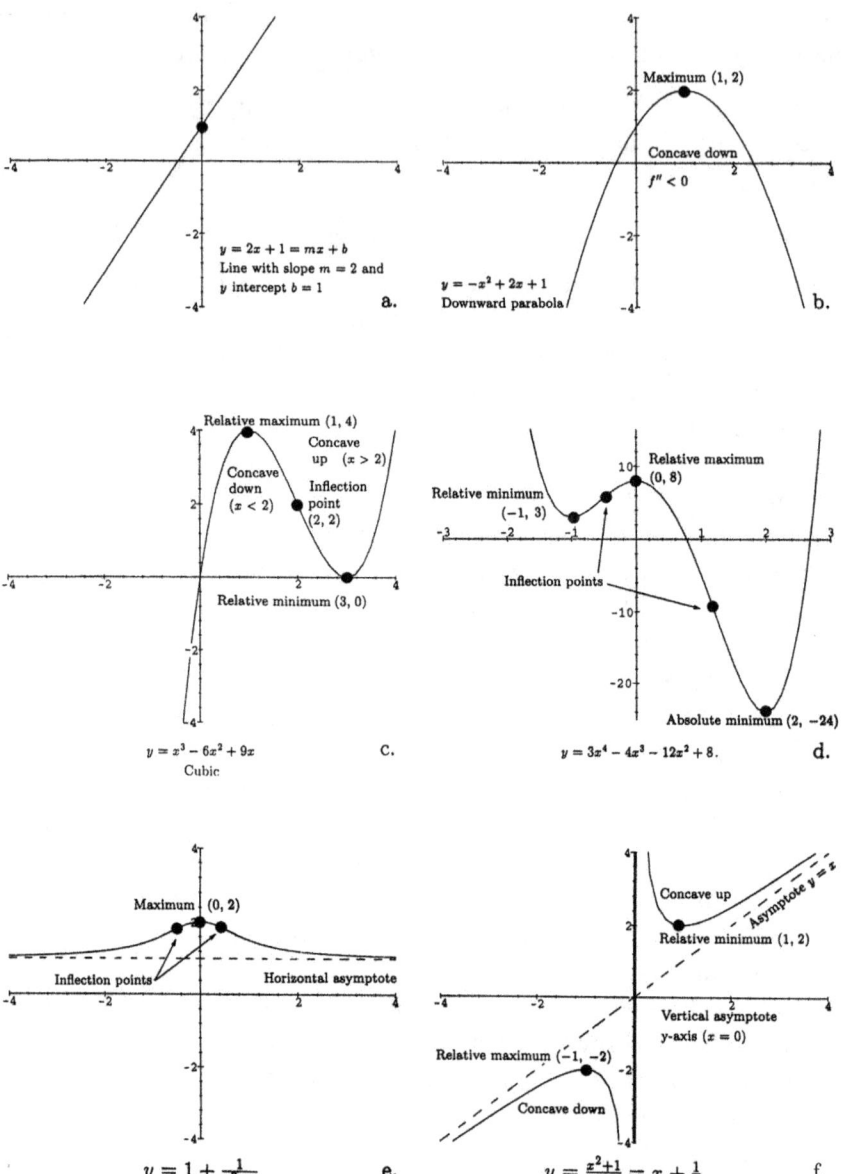

Figure 11.3. Important features of graphs.

Figure 11.3(f) graphs a function with a fraction that blows up as $x \to 0$, and the line $x = 0$ is a vertical asymptote. Since $\lim_{x \to 0+} y = +\infty$, the graph follows the asymptote upwards from the right. Since $\lim_{x \to 0-} y = -\infty$, the graph follows the asymptote downwards from the left. For x very large positive or negative, $y \approx x$, and the graph approaches the asymptote $y = x$.

Now it is time for a harder example that combines some of these features.

Example 11.1. Sketch the graph of

$$y = f(x) = \frac{x^2 - 7x + 12}{x - 2}.$$

Steps (a) and (b). First, factor

$$y = f(x) = \frac{(x-3)(x-4)}{x-2}.$$

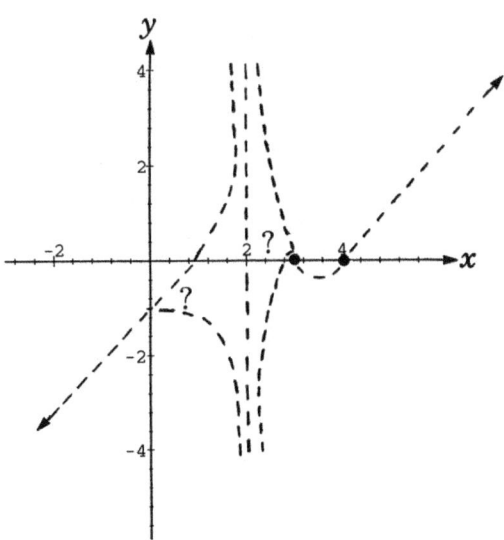

Figure 11.4. A first attempt at the graph of $y = f(x) = \frac{x^2-7x+12}{x-2} = \frac{(x-3)(x-4)}{x-2}$, exhibiting the zeros of f and the behavior as $x \to \pm\infty$.

The function f is defined on all of the real line, except that it blows up at 2. As $x \to +\infty$, $y \approx x^2/x \approx x \to +\infty$; as $x \to -\infty$, $y \approx x^2/x \to -\infty$. The function is 0 at 3 and 4. See Figure 11.4.

Let's look at x near 2 more carefully and consider $\lim_{x \to 2} f(x)$. For x a bit bigger than 2, the numerator is positive (about $(-1)(-2) = 2$), the denominator is small and positive, so $\lim_{x \to 2+} f(x) = +\infty$. For x a bit smaller than 2, the numerator is still positive, but the denominator is small and negative, so $\lim_{x \to 2-} f(x) = -\infty$. Hence the graph approaches the vertical asymptote $x = 2$ downward from the left and upward from the right. Of course y could not have gone to $+\infty$ from the left of 2 — or it would have been 0 somewhere left of 2.

Consider $y(0) = -6$, $y(1) = -6$. There is no room for such negative values in our picture. For a better graph, we would need to change the scale. See Figure 11.5. (Finding the right scale is often a matter of trial and error. Be willing to redraw the axes as we will again in Figure 11.6; it does not take long.)

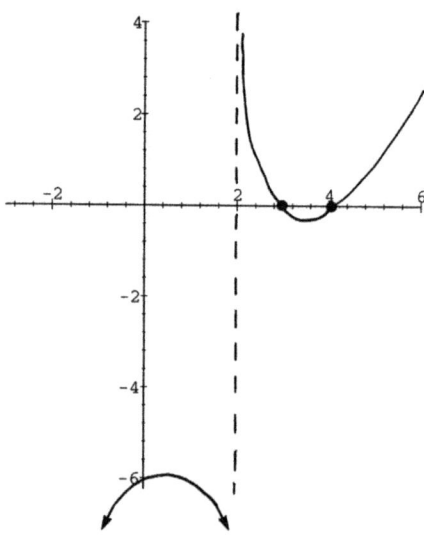

Figure 11.5. A better sketch of the graph of $y = f(x) = \frac{x^2 - 7x + 12}{x - 2} = \frac{(x-3)(x-4)}{x-2}$ from consideration of what happens for x near 2.

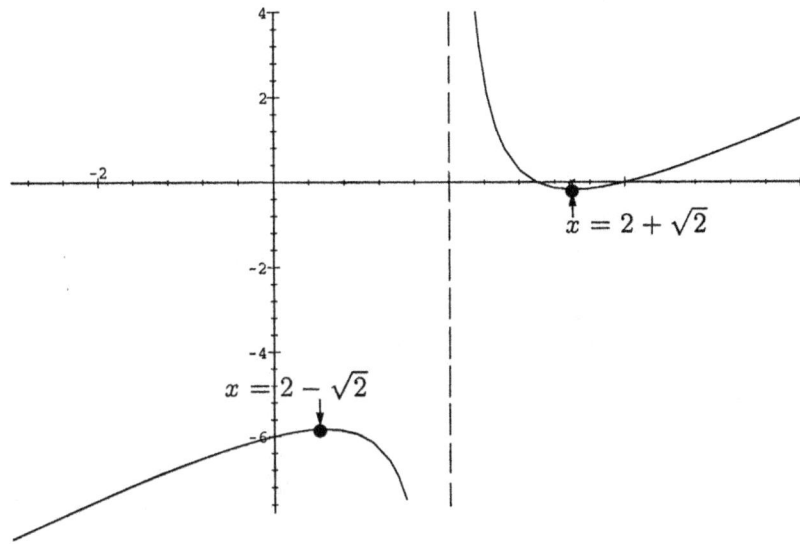

Figure 11.6. A still better sketch of the graph of $y = f(x) = \frac{x^2-7x+12}{x-2} = \frac{(x-3)(x-4)}{x-2}$ showing the relative extrema where $f'(x) = 0$.

Step (c). There are two obvious relative extrema. To locate them, use the derivative:

$$y' = \frac{(2x-7)(x-2) - (x^2 - 7x + 12)}{(x-2)^2}$$

$$= \frac{x^2 - 4x + 2}{(x-2)^2}.$$

The derivative y' is positive as $x \to \pm\infty$ (good). The derivative y' is zero when

$$x = \frac{4 \pm \sqrt{4^2 - 4 \cdot 2}}{2} = 2 \pm \sqrt{2} \approx .6, \; 3.4.$$

See Figure 11.6.

Step (d). Finally check the second derivative. We expect $f''(x) < 0$ when $x < 2$ and $f''(x) > 0$ when $x > 2$.

$$y'' = \frac{(2x-4)(x-2)^2 - (x^2 - 4x + 2) \cdot 2(x-2)}{(x-2)^4}$$

$$= \frac{4}{(x-2)^3},$$

which behaves as expected. Note how the second derivative is negative at the local maximum and positive at the local minimum.

11.2. Higher Order Derivatives

Taking the derivative of $y = f(x)$ n times yields the nth derivative:

$$y^{(n)} = \frac{d^n y}{dx^n} = f^{(n)}(x).$$

We will use parentheses on the superscript (n) to distinguish the nth derivative $y^{(n)}$ from the nth power y^n.

Exercises 11

Given the mileage $f(t)$, compute the velocity and acceleration.

1. $f(t) = -16t^2 + 12t$

2. $f(t) = e^t \sin t$

3. $f(t) = t^2 e^t \sin t$

4. $f(t) = t^2 e^t (\sin t) \ln(t^2 + 1)$

Compute the first four derivatives.

5. $y = x^4 e^x$

6. $y = x^4 \sin x$

7. $y = \frac{1}{36} x^3 (6 \ln x - 11)$

For the following graphs $y = f(x)$, where are $f'(x)$ and $f''(x)$ positive and negative?

8.

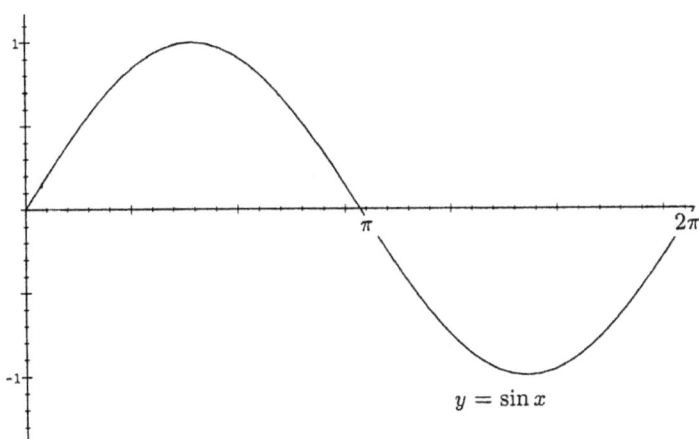

$y = \sin x$

9.

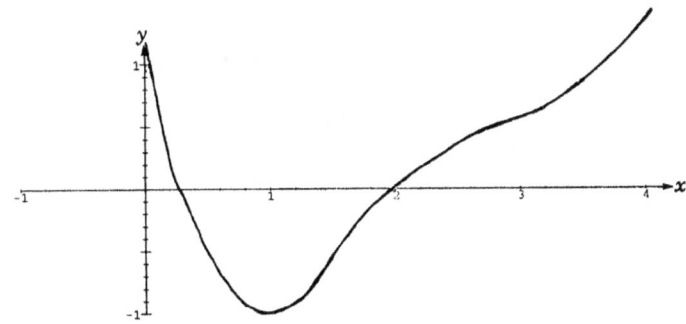

Sketch the graphs of $y = f(x)$.

10. $y = 3x + 2$

11. $4x + 2y = 6$ (Solve for y first.)

12. $y = x^2 - 4x + 2$

13. $y = -\frac{1}{2}x^2 + x - 1$

Label absolute and relative maxima and minima and points of inflection.

14.

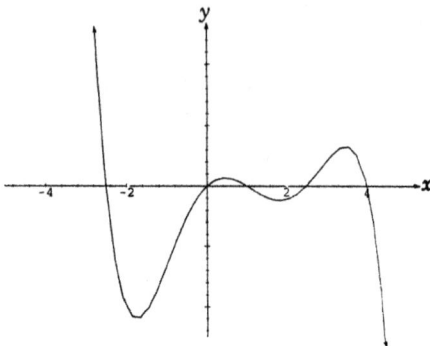

15.

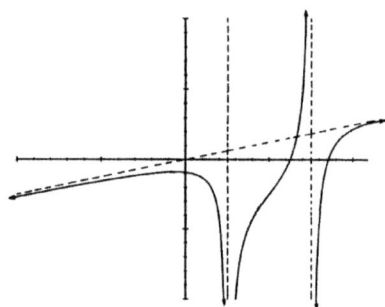

16.

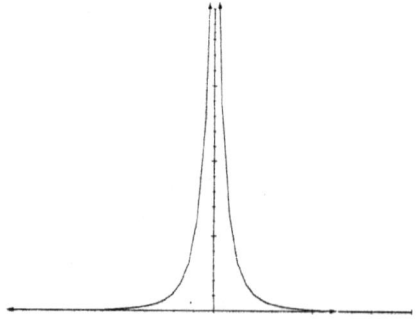

17. Label where $y' > 0$ on this graph $y = f(x)$.

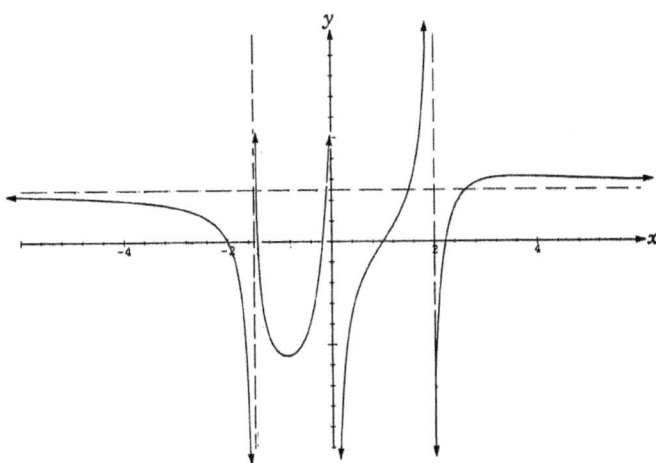

Sketch the graphs of $y = f(x)$. Label relative maxima and minima and inflection points.

18. $y = \dfrac{x - 2}{x - 5}$ (Note the horizontal asymptote $y = 1$, approached as $x \to \pm\infty$.)

19. $y = x^3 - 8x^2 + 5x + 1$

20. $y = x^4 - 2x^2 + 1$

21. $y = x - 4x^{-1}$ (Note the asymptotes $x = 0$ and $y = x$.)

22. $y = \dfrac{x^2}{x - 4}$

23.   a. $y = \sqrt{x}$

      b. $y = \sqrt[3]{x}$

      c. $y = x^{\frac{2}{3}}$

24. $y = xe^x$

25. $y = \dfrac{\sqrt{x}}{x^2 + 1}$

Sketch the following graphs (from old examinations):

26. $y = x^3 - 6x^2 + 9x$

27. $y = \dfrac{x^2 + 3x + 1}{x^2 + 2x + 1}$

28. $y = x + \dfrac{1}{x - 2}$, given that $y' = \dfrac{(x-1)(x-3)}{(x-2)^2}$, $y'' = \dfrac{2}{(x-2)^3}$.

12

Antidifferentiation

An old car travels down the interstate for three hours, slowly gaining speed. Suppose the speedometer at time t shows a velocity of

$$\text{velocity } = v(t) = 40t.$$

What is the reading $f(t)$ of the odometer at each time t ? Since the velocity v is the derivative of the odometer mileage reading f, f must be an *antiderivative* of v. Possibilities are $f(t) = 20t^2$ (take the derivative and you get $40t$, sure enough), plus any constant (since the derivative of a constant is 0):

$$f(t) = 20t^2 + C.$$

Plugging in $t = 0$, you see that the constant C here must be the odometer reading $f(0)$ at time $t = 0$. Knowing $f(0)$ and the velocity at all times determines the odometer reading $f(t)$ at all times. All possible antiderivatives are given by $f(t) = 20t^2 + C$. (It is actually not easy to prove mathematically that there are not any others, and that is the reason for all the fuss about the "Mean Value Theorem" in other books.)

In general, antidifferentiation introduces a constant C , which can only be determined by some initial conditions or other given information. When such information is not given, you need to leave the +C.

Table 12.1. Antiderivatives of some important functions.

Function	Antiderivative
0	C
a	$ax + C$
x	$\frac{1}{2}x^2 + C$
x^r	$\frac{x^{r+1}}{r+1} + C \quad (r \neq -1)$
x^{-1}	$\ln|x| + C \quad (x \neq 0)$
$\cos ax$	$\frac{1}{a}\sin ax + C$
$\sin ax$	$-\frac{1}{a}\cos ax + C$
e^{ax}	$\frac{1}{a}e^{ax} + C$

Table 12.1 gives antiderivatives of some important functions. The table can be verified by just checking that differentiation yields the original function back. For example,

$$\frac{d}{dx}\left(\frac{x^{r+1}}{r+1} + C\right) = x^r.$$

Note that for $x > 0$, $\ln x$ is defined and

$$\frac{d}{dx}\ln x = x^{-1}.$$

For $x < 0$, $\ln(-x)$ is defined, and

$$\frac{d}{dx}\ln(-x) = -(-x)^{-1} = x^{-1}.$$

In either case,

$$\frac{d}{dx}\ln|x| = x^{-1}.$$

Examples 12.1. The antiderivative of $5x^2 + 3x$ is

$$5\frac{x^3}{3} + 3\frac{x^2}{2} + C = \frac{5}{3}x^3 + \frac{3}{2}x^2 + C.$$

The antiderivative of $\dfrac{x\cos 5x - 6}{x} = \cos 5x - 6x^{-1}$ is

$$\frac{1}{5}\sin 5x - 6\ln|x| + C.$$

We do not yet know how to find the antiderivative of $\sin^2 x$:

$$f(x) = \frac{1}{3}\sin^3 x$$

does not work, because

$$f'(x) = \frac{1}{3}\,3\,(\sin^2 x)(\cos x) \neq \sin^2 x.$$

Similarly, we do not yet know the antiderivatives of $\sin x^2$ or e^{x^2}. The antiderivative of $(e^x)^2 = e^{2x}$ is $\frac{1}{2}e^{2x}$.

Falling Objects have a downward acceleration due to gravity, about -32 feet/sec^2 near the Earth's surface. The height $f(t)$ of a falling object may be found by antidifferentiating the constant -32, once to get the velocity $v(t) = -32t + v_0$, again to get the height

$$f(t) = -16t^2 + v_0 t + f_0.$$

The constants of integration v_0 and f_0 are the initial velocity and initial height. Exercises 23–25 provide some examples.

Exercises 12

Find all antiderivatives.

1. x^5

2. $x^3 + 7x$

3. $9x^3 - x + 3$

4. $2x^{11} - 9x^8 - 1$

5. x

6. 10

7. 1

8. 0

9. $\sqrt{x}$

10. $\sqrt[3]{x}$

11. $\dfrac{1}{x^{10}}$

12. $\dfrac{1}{\sqrt{x}}$

13. $\dfrac{5}{x}$

14.   a. $\cos 3x + \sin 3x$

   b. $5 \cos \frac{1}{2}x - 7 \sin \frac{1}{2}x$

15. $\cos x^2$

16. e^{7x}

17. $e^{7/x}$

Given the acceleration $a(t)$ and further information, find the odometer reading $f(t)$.

18. $a(t) = 6t,\ v(0) = -1,\ f(0) = 0$

19. $a(t) = 2,\ v(0) = -1,\ f(0) = 16$

20. $a(t) = -\sin t,\ v(0) = 1, f(0) = 0$

21. $a(t) = -\sin t,\ v(0) = 2,\ f(0) = 0$

22. $a(t) = e^{t/10},\ v(0) = 10, f(0) = 0$

23. Find the height $f(t)$ of a cat that steps off a 15-foot ladder. (Use $a = -32, v_0 = 0, f_0 = 15$).

24.   a. Find an equation for the height $f(t)$ of a penny dropped from the top of the Empire State Building, which is 1250 feet tall. (Use $a = -32$, $v_0 = 0$, $f_0 = 1250$).

    b. When will it hit the ground?

    c. How fast is it going when it hits the ground?

    (DO NOT TRY THIS!)

25. (From an old final.) A ball is thrown straight up from the top of a 200 foot building with an initial velocity of 40 feet/second. Acceleration due to gravity is -32 feet/sec^2.

    a. Find a formula for the height of the ball at time t.

    b. How high does the ball rise before it begins to fall?

    c. When does it hit the ground (assuming it does not fall back onto the roof of the building)?

13

Differentiability and Continuity

Many functions fail to be differentiable at one or more points: the graph fails to have a definite tangent line at these points, perhaps by having corners or, worse, jumps in the graph. For example, the function $f(x) = |x|$ fails to be differentiable at 0. (See Figure 13.1.) The function $f(x) = |\sin x|$ fails to be differentiable at every multiple of π. The function $f(x) = [x] =$ the greatest integer $n \leq x$ fails to be differentiable at every integer. Some functions are actually differentiable nowhere.

Technically, failure to be differentiable means that the limit

$$f'(a) = \lim_{x \to a} \frac{f(x) - f(a)}{x - a}$$

does not exist; *i.e.*, as $x \to a$, the ratio $\frac{f(x)-f(a)}{x-a}$ does not home in on a single value. For example, for $f(x) = |x|$,

$$f'(0) = \lim_{x \to 0} \frac{|x|}{x}$$

does not exist. If x is positive, $|x|/x = 1$, but if x is negative, $|x|/x = -1$. As $x \to 0$, you get ± 1, depending on whether x is

111

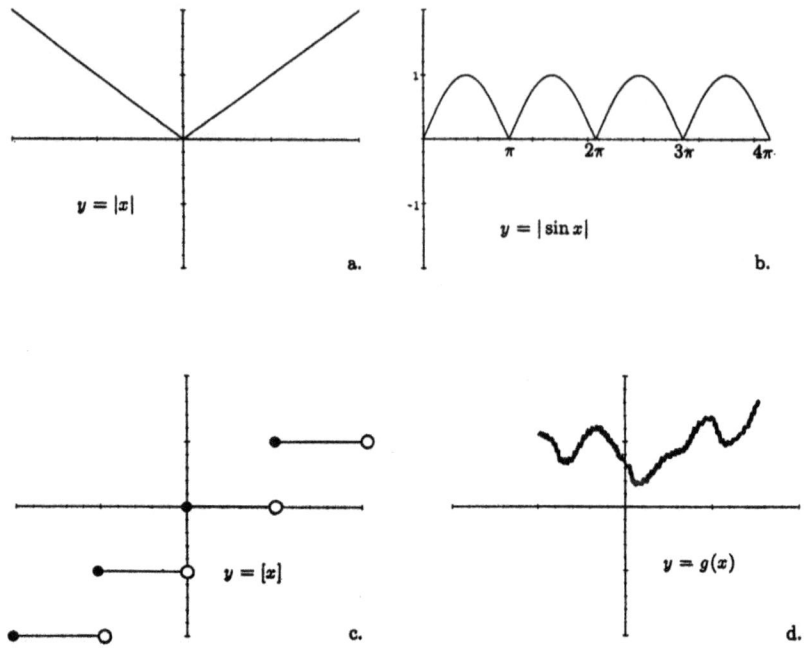

Figure 13.1. Examples of functions which fail to be differentiable at one or more points.

positive or negative. In worse examples, you might pick up not just two, but infinitely many limiting values as $x \to a$.

13.1. Higher Differentiability

Even if $f(x)$ is differentiable, $f'(x)$ might not be. For example, if $f(x) = x^{5/3}$, then $f'(x) = \frac{5}{3}x^{2/3}$, and $f''(x) = \frac{10}{9}x^{-1/3}$, which blows up at 0. See Figure 13.2.

Even if $f(x)$ is twice differentiable, it might not be three times differentiable, and so on. Now maybe you can begin to appreciate how nice it is of many of our familiar functions to have derivatives of all orders at every point.

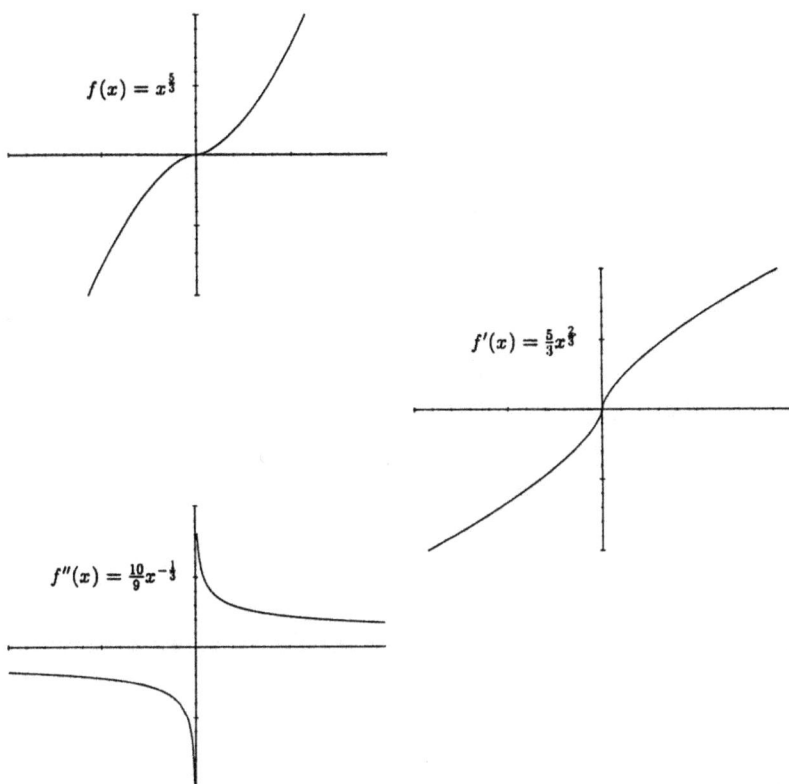

Figure 13.2. The function $f(x) = x^{5/3}$ is differentiable but not twice differentiable at 0. Its second derivative blows up at 0.

13.2. Continuity

A function is called *continuous* if its graph has no jumps, *i.e.*, if its graph can be drawn without lifting the pencil from the paper. Of the four examples of Figure 13.1, all except (c) are continuous. Technically, we say $f(x)$ is *continuous* at a if

$$\lim_{x \to a} f(x) = f(a).$$

This fails for $y = [x]$ at 0 because $\lim_{x \to 0}[x]$ does not exist (you get -1 from the left and 0 from the right). For the discontinuous

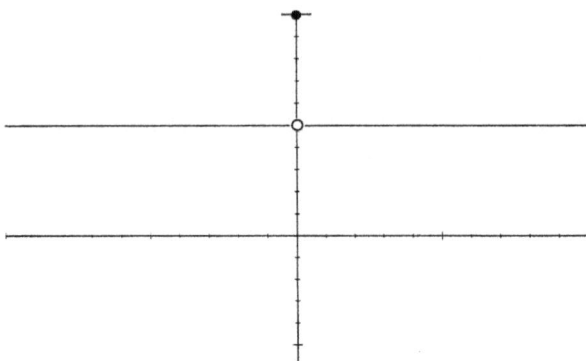

Figure 13.3. This function is not continuous at 0.

function graphed in Figure 13.3,

$$\lim_{x\to 0} f(x) = 1 \neq f(0) = 2.$$

Every differentiable function must be continuous. But a continuous function need not be differentiable, as shown by Examples a, b, d of Figure 13.1.

Exercises 13

Is each of the following functions differentiable and continuous at all points? If not, where do differentiability and continuity fail?

1.

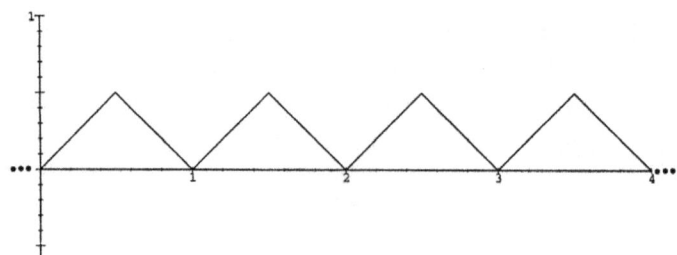

2.

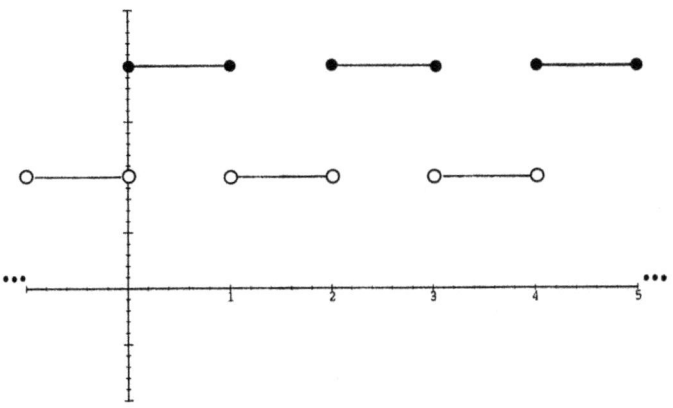

3.

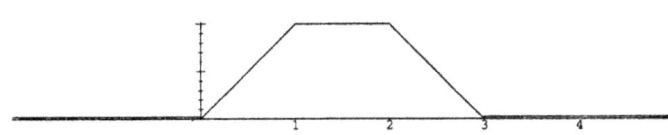

4.

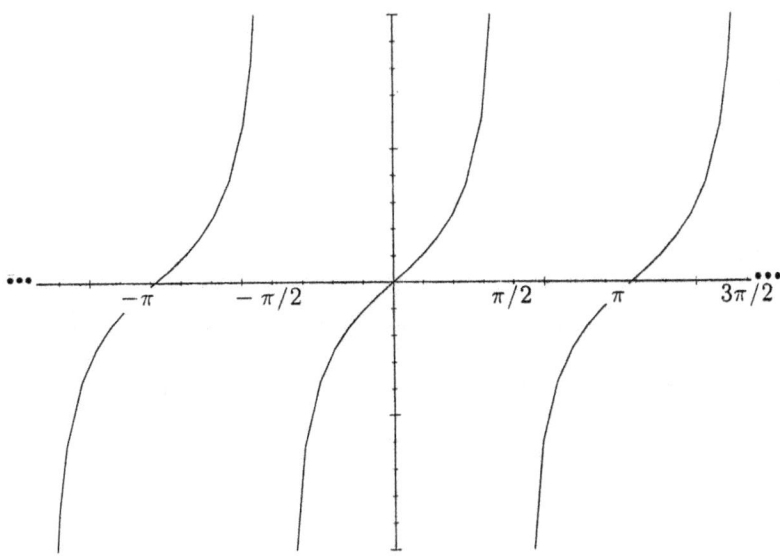

5. $f(x) = |\cos x|$

6. $f(x) = |x^2 - 3x + 2|$

7. $f(x) = \sqrt{|x|}$

8. $f(x) = [x^2]$

9. $f(x) =$ the distance from x to the nearest integer

10. $f(x) = \dfrac{x}{\pi}$

11. $f(x) = x^\pi$

12. $f(x) = x^2 - 2x + 8$

13. $f(x) = \cos(x^{10} - 9x^3 + 7)$

14.   a. $f(x) = \dfrac{1}{x}$

      b. $f(x) = \dfrac{\pi}{1 - x}$

      c. $f(x) = \dfrac{x}{1 - \pi}$

15. $f(x) = \dfrac{1}{x^2 + 4x - 5}$

16. $f(x) = x + |x|$

17. $f(x) = \sin x + \cos x$

18. $f(x) = x \sin x$

19. $f(x) = \sin \dfrac{1}{x}$

20. Are all polynomials continuous and differentiable everywhere?

21. If $f(x)$ and $g(x)$ are continuous and differentiable everywhere, is $f(x) + g(x)$? $f(x)g(x)$? $f(x)/g(x)$? $f(g(x))$?

14

Review

Formulas by Sections

Section 1. Definition of Derivative.

- $f'(x) = \lim\limits_{\Delta x \to 0} \dfrac{f(x + \Delta x) - f(x)}{\Delta x}$

Section 3. Product Rules.

- $(fg)' = f'g + fg'$
- $(fgh)' = f'gh + fg'h + fgh'$

  Quotient Rule.

- $\left(\dfrac{f}{g}\right)' = \dfrac{f'g - fg'}{g^2}$

Section 5. Power Rule.

- $(x^r)' = rx^{r-1}$
- $(u^r)' = ru^{r-1}u'$

Section 6. Sines and Cosines.

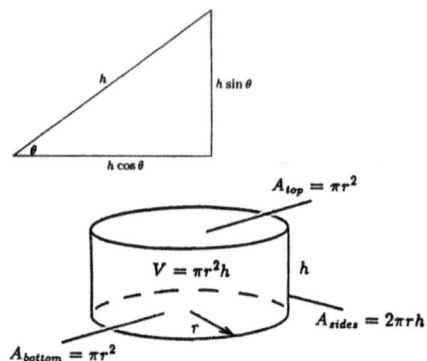

- $\sin^2\theta + \cos^2\theta = 1$
- $2\cos\theta\sin\theta = \sin 2\theta$
- $(\sin x)' = \cos x$
- $(\sin u)' = (\cos u)u'$
- $(\cos x)' = -\sin x$
- $(\cos u)' = (-\sin u)u'$

Section 9. Exponentials and Logarithms.

- $\ln xy = \ln x + \ln y$
- $\ln \dfrac{x}{y} = \ln x - \ln y$
- $\ln \dfrac{1}{x} = -\ln x$
- $\ln x^p = p\ln x$
- $\log_a x = \dfrac{\ln x}{\ln a}$
- $(e^x)' = e^x$
- $(e^u)' = e^u u'$
- $(a^x)' = (\ln a)a^x$
- $(a^u)' = (\ln a)a^u u'$
- $(\ln x)' = \dfrac{1}{x}$
- $(\ln u)' = \dfrac{u'}{u}$

Section 10. Exponential Growth.

- Population and money: $y = y_0 e^{kt}$
- Radioactive material: $y = y_0(\frac{1}{2})^{t/h}$

  y amount at time t, y_0 initial amount, k growth rate, h half-life

Exercises 14

Compute the derivatives.

1. $\sin 5x$

2. $\cos \dfrac{\pi - x}{2}$

3. $\sin(x^3 - x^2)$

4. $e^{\frac{x}{50}+7}$

5. 2^{10x}

6. $x^2 e^{4x+3}$

7. $\tan x = \dfrac{\sin x}{\cos x}$

8. $\tan 2x^2$

9. $\dfrac{e^x - e^{-x}}{e^x + e^{-x}}$

10. $\dfrac{(x-5)^5 (x-4)^4 (x-3)^3 (e^{x^2})}{x^{12}}$

11. $\sin^4 2x \sin^5 3x$

12. $x^{\sin x}$

13. Calculate the derivatives.

    (a) $\dfrac{d}{dx}(\sin(3x^2 + 2))$

    (b) $\dfrac{d}{dx}(\cos(\ln 5x))$

(c) $\dfrac{d}{dx}((\sin(x+2))\ln 3x)$

(d) $\dfrac{d}{dx}(3x^{\ln x})$

(e) $\dfrac{d}{dx}(\sin^2 x + \cos^2 x)$

(f) $\dfrac{d}{dx}\left(\dfrac{\cos x}{x^2+3}\right)$

(g) $\dfrac{d}{dx}(x\cos 3)$

(h) $\dfrac{d}{dx}(\cos^2(12x^3 - 8x))$

(i) $\dfrac{d}{dx}(\cos(\sin x))$

(j) $\dfrac{d}{dx}(e^{\cos^3(x^2+1)})$

(k) $\dfrac{d}{dx}(\ln e^3)$

(l) $\dfrac{d}{dx}(x^2 \cos x)$

(m) $\dfrac{d}{dx}\left(\dfrac{e^{2x}}{x^2}\right)$

(n) $\dfrac{d}{dx}(100 \cdot 2^x)$

(o) $\dfrac{d}{dx}(4^{3x})$

(p) $\dfrac{d}{dx}(\ln 3x^2)$

14. An open rectangular box with square ends is to hold 6400 cu. ft. and is to be constructed at a cost of \$.75 per sq. ft. for the base and \$.25 per sq. ft. for the sides. Find the most economical dimensions.

15. Let $f(x) = \dfrac{x^2 + x + 1}{x^2 + 1}$. Then

$$f'(x) = \frac{1 - x^2}{(x^2 + 1)^2},$$

and

$$f''(x) = \frac{2x(x^2 - 3)}{(x^2 + 1)^3}.$$

Graph $f(x)$.

16. Use the *definition* of the derivative to find $\dfrac{d}{dx}\left(\dfrac{1}{x}\right)$.

17. Compute the following derivatives. (You need not simplify your answers completely.)

(a) $\dfrac{d}{dx}\left(\sqrt[3]{x^2 + 2x + 1}\right)$

(b) $\dfrac{d}{dx}(\ln(\sin x))$

(c) $\dfrac{d}{dx}(e^x \cdot \cos x)$

(d) $\dfrac{d}{dt}\left(\dfrac{1}{\ln t}\right)$

(e) $\dfrac{d}{dt}(\cos t^2)$

(f) $\dfrac{d}{dx}\left(x^{e^x}\right)$

18. Evaluate: $e^{4\ln 2}$, $\cos^2(5\pi/4)$, $\ln e^{\ln 1}$, $\ln e^{\cos^2 x}$.

19. A particle moves along a straight line with acceleration $a(t) = 2 - t^2$. If $s(0) = -1$ and $v(0) = 1$, find $s(t)$.

20. Find the equation of the tangent line to $x \cos y + x^2 y = 2$ at $(2, 0)$.

21. A population of bacteria is known to have a population of $P(t) = 50e^{t/10}$ at time t (hours). Is the population growing? Why? When is the population 500? How fast is it growing when there are 70?

22. Find the slope of the line tangent to $x^4 + xy + y^4 = 19$ at the point (1,2).

23. Find the equation of the line tangent to the graph of $y = x^2 e^{-x^2}$ at the maximum.

24. Sketch the graph of $f(x) = e^x + e^{-x}$.

25. Sketch the graph of $f(x) = e^{-x}$.

26. When Carla the Human Cannonball is fired vertically out of her cannon, her height above the ground is given by $s(t) = 256 + 96t - 16t^2$, where t is the number of seconds since the cannon was fired and $s(t)$ is in feet. Find the highest point she reaches.

27. Compute the antiderivatives.

    a. $3x^3 + 2x^2$

    b. $\sqrt{x} + \sqrt[3]{x}$

    c. $\sin \dfrac{x}{100}$

    d. $\sin \sqrt{x}$

    e. $e^{x/100}$

    f. e^{x^2}

    g. $2xe^{x^2}$

28. Show that the Quotient Rule can be derived directly from the Product Rule (as opposed to using the definition of the derivative) by applying the Product Rule to $\dfrac{f(x)}{g(x)} = fg^{-1}$.

29. Using the circle $x^2 + y^2 = 1$ centered at the origin, show that two points directly opposite on the circle have tangent lines with the same slope.

30. Find y' if

    a. $x^2y^2 + xy + 7 = 0$,

    b. $(\cos x)(\sin y) = \dfrac{1}{2}$,

    c. $(x + y)^2 = x^4 + y^4 + xy$.

31. Derive the formula for $(x^4)'$ from the definition of the derivative.

32. Simplify or say "cannot be simplified:"
$$27^{2/3}, \ 27^{-4/3}, \ 8^{-3}, (\sqrt[6]{x})^{12}, \ (x^{-3}y^{-6})^{-1/3}, \ (x^6 + y^{12})^{1/3}.$$

33. Find the equation of the line tangent to the graph $y = \sqrt{x^5}$ at $x = 4$.

34. Find $y'(1)$ if
$$y + \sqrt[3]{x^4/5} + \cos^3 x^5 + \frac{2xy}{1 + x^2} = 2.$$

Find the absolute extrema:

35. $y = \dfrac{1}{3}x^3 - \dfrac{1}{2}x^2 - 2x + 3$ for $x \geq 0$;

36. $y = 2x^3 - 15x^2 + 6x - 1$ for $0 \leq x \leq 5$;

37. $y = x^3 + x^2 + x + 1$ for $x \leq 0$;

38. $y = \dfrac{x^3}{27 + x^4}$.

Find the derivatives (from an old final):

39. $f(w) = \dfrac{w}{w-1} + \dfrac{w+1}{3w}$;

40. $y = \sqrt{\cos \sqrt{x}}$;

41. $g(x) = x\sqrt{\ln x}$;

42. $f(t) = \ln(te^t)$;

43. $y = \sqrt[3]{\cos^2 x + \sin x^2}$.

Part II

The Integral

15

Area and the Riemann Integral

The ancient Greeks knew the area of a right triangle is $\frac{1}{2}bh$ and used a limit argument to deduce that the area of a disc (the area inside a circle) is $\frac{1}{2}Cr = \pi r^2$, where C is the circumference. See Figure 15.1.

Using a more complicated argument, they decided that the volume of a ball is $\frac{4}{3}\pi r^3$. They never knew that the volume of a 4-dimensional ball is $\frac{1}{2}\pi^2 r^4$.

Integral calculus provides a systematic method for computing areas and volumes. We begin by asking for the area under the graph of a function, between $x = a$ and $x = b$, such as the area under $y = x^2$ for x between 0 and 1, as in Figure 15.2.

The method is to cut the x-interval from a to b into little subintervals and estimate the area over each subinterval by say a circumscribed rectangle. The length Δx of the subinterval becomes the width of the rectangle. The height of each rectangle is $f(x)$, where x is the place in the subinterval where f is biggest. See Figure 15.3. The area of each rectangle is $f(x)\Delta x$. The estimate given by the sum of the areas of the rectangles is the sum of all the $f(x)\Delta x$, abbreviated $\sum f(x)\Delta x$. The capital Greek letter Sigma ($\sum$) is used to

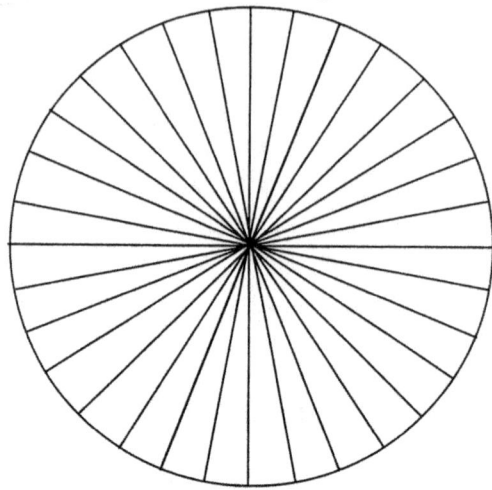

Figure 15.1. A disc is composed of infinitely many infinitesimal triangles of height r and total base C for a total area of $\frac{1}{2}Cr = \pi r^2$.

denote "sum of all." This area-approximating sum is called a *Riemann sum*.

The smaller the subintervals, the more accurate the estimate. The exact area comes in the limit as $\Delta x \to 0$:

$$\lim_{\Delta x \to 0} \sum f(x)\Delta x.$$

This limit is so important that it is given a name, the *Riemann integral* or *definite integral*, and its own symbol,

$$\int_a^b f(x)dx = \lim_{\Delta x \to 0} \sum f(x)\Delta x. \qquad (15.1)$$

The Riemann integral is thus defined as a limit of Riemann sums. It gives the area under $y = f(x)$ for $a \le x \le b$.

Actually it is not necessary that the x in the $f(x)$ on the right-hand side of Eq. 15.1 be chosen where f is biggest; x can be chosen anywhere in the subinterval. This will make the approximating Riemann sums smaller, but the difference gets smaller as the intervals get shorter, and the limit gives exactly the same value for the area under the graph of any continuous function.

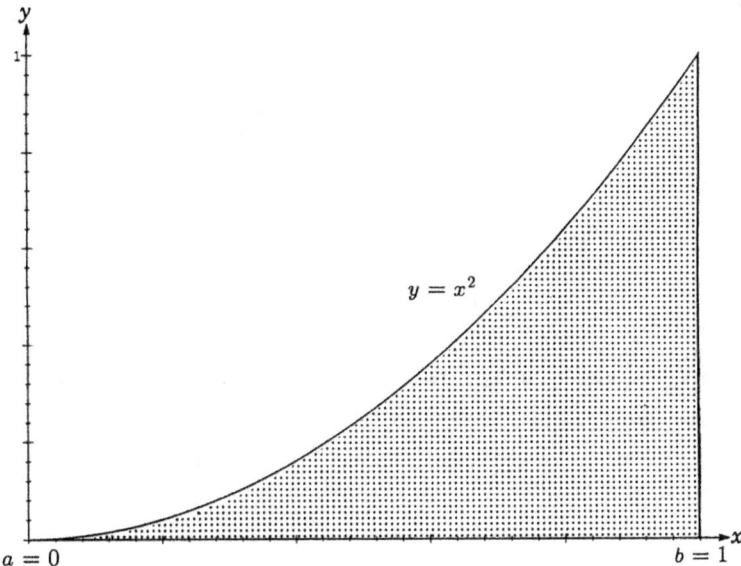

Figure 15.2. The area under x^2 looks like something between 1/2 and 1/4.

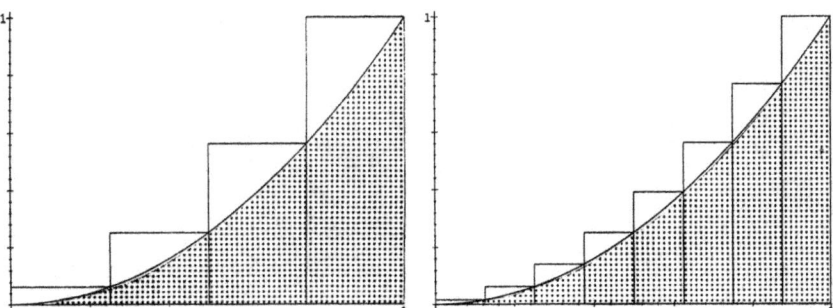

Figure 15.3. The smaller the subintervals, the better the circumscribed rectangles estimate the area under the curve.

If the interval from a to b is divided up into n equal subintervals, each has length $\Delta x = \frac{b-a}{n}$, and $\Delta x \to 0$ as $n \to \infty$. We can get to an x in the kth subinterval by starting at a and taking k steps, arriving at

$$x = a + k\frac{b-a}{n}.$$

Therefore we could write more definitely

$$\int_a^b f(x)dx = \lim_{n \to \infty} \sum_{k=1}^n f\left(a + k\frac{b-a}{n}\right)\frac{b-a}{n}.$$

We have just rewritten the Riemann sum; it still adds up a contribution from each subinterval of a value of f times the Δx.

Let us try to carry out this process to compute the area under $y = x^2$ for $0 \le x \le 1$:

$$\int_0^1 x^2 dx = \lim_{\Delta x \to 0} \sum x^2 \cdot \Delta x.$$

Divide the interval into n subintervals of length

$$\Delta x = \frac{b-a}{n} = \frac{1}{n}.$$

In the kth subinterval x goes from $\frac{k-1}{n}$ to $\frac{k}{n}$. The circumscribed rectangle has height $\left(\frac{k}{n}\right)^2$. The Riemann sum of the areas of the rectangles is

$$\sum x^2 \Delta x = \sum_{k=1}^n \left(\frac{k}{n}\right)^2 \left(\frac{1}{n}\right) = \frac{1}{n^3}\sum_{k=1}^n k^2.$$

There is a formula for the sum of the first n squares:

$$\sum_{k=1}^n k^2 = \frac{n(n+\frac{1}{2})(n+1)}{3}.$$

(Check that for $n = 1$ it gives $1^2 = 1$ and for $n = 2$ it gives $1^2 + 2^2 = 5$.) Therefore as $n \to \infty$ and $\Delta x \to 0$, we obtain

$$A = \int_0^1 x^2 dx = \lim_{\Delta x \to 0} \sum x^2 \cdot \Delta x$$

$$= \lim_{n\to\infty} \frac{1}{n^3} \cdot \frac{n(n+\frac{1}{2})(n+1)}{3}$$

$$= \lim_{n\to\infty} \frac{n}{n} \cdot \frac{n+\frac{1}{2}}{n} \cdot \frac{n+1}{n} \cdot \frac{1}{3}$$

$$= \lim_{n\to\infty} 1(1+\frac{1}{2n})(1+\frac{1}{n})\frac{1}{3} = \frac{1}{3}.$$

The area is exactly $1/3$.

Notice that this process involves some formula, here the formula for the sum of the first n squares.

Exercises 15

1. Following the example in the text, compute the area under $y = x$ for $0 \le x \le 1$, and check that you get the expected answer $1/2$. For the Riemann sum you should get

$$\sum_{k=1}^{n} \left(\frac{k}{n}\right)\left(\frac{1}{n}\right) = \frac{1}{n^2} \cdot \frac{n(n+1)}{2}.$$

You will need to use the formula that the sum of the first n integers is $\frac{n(n+1)}{2}$ (the number n of integers times their average $\frac{n+1}{2}$).

2. Compute $\int_2^4 x\,dx$. For the Riemann sum you should get

$$\sum_{k=1}^{n} \left(2 + \frac{2k}{n}\right) \cdot \left(\frac{2}{n}\right) = \frac{4}{n}\sum_{k=1}^{n} 1 + \frac{4}{n^2}\sum_{k=1}^{n} k$$

$$= \frac{4}{n} \cdot n + \frac{4}{n^2} \cdot \frac{n(n+1)}{2}.$$

Check your answer against the formula for the area of a trapezoid: base times average height.

3. Memorize: We approximate an area by a *Riemann sum* of areas of n skinny rectangles of height $f(x)$ and width $\Delta x = \frac{b-a}{n}$. The exact area is given by the limit as $n \to \infty$, called the *Riemann integral*.

16

The Fundamental Theorem of Calculus

Areas or definite integrals are hard to compute directly as limits of Riemann sums. The Fundamental Theorem of Calculus tells us that definite integrals can be easily computed by *antidifferentiating*. Our work on derivatives comes from out of the blue to save us!

16.1. The Fundamental Theorem of Calculus

To compute $\int_a^b f(x)\,dx$ for any continuous function f on $[a,b]$, take any antiderivative F of f. Then

$$\int_a^b f(x)\,dx = F(b) - F(a) = F(x)]_a^b.$$

The final symbol is just a shorthand for $F(b) - F(a)$.

   Example 16.1. Compute the area under $y = x^2$ for $0 \le x \le 1$. Here $f(x) = x^2$, and one antiderivative is $F(x) = \frac{1}{3}x^3$. (Since we can choose any antiderivative, $\frac{1}{3}x^3 + C$, we choose $C = 0$. (It would cancel out in $F(b) - F(a)$ anyway.) Therefore the area is given by

$$\int_a^b x^2 \, dx = \left.\frac{1}{3}x^3\right]_0^1 = \frac{1}{3}\cdot 1^3 - \frac{1}{3}\cdot 0^3 = \frac{1}{3}.$$

Example 16.2. Find the area under one arch of the function $y = \sin 5x$. See Figure 16.1.

Solution. When $\theta = 0$, $\sin\theta$ is 0 and returns to 0 when θ reaches π. Therefore one arch corresponds to $0 \le 5x \le \pi$, i.e., $0 \le x \le \frac{\pi}{5}$. The integrand $f(x) = \sin 5x$, and one tive is $-\frac{1}{5}\cos 5x$. Therefore the area is given by

$$\int_0^{\frac{\pi}{5}} \sin 5x \, dx = \left.-\frac{1}{5}\cos 5x\right]_0^{\frac{\pi}{5}} = -\frac{1}{5}\cos\pi - -\frac{1}{5}\cos 0 = \frac{1}{5} + \frac{1}{5} = \frac{2}{5}.$$

You must be wondering what antidifferentiation has to do with area or the definite integral. The key idea behind this relationship as given by the Fundamental Theorem is actually a simple one.

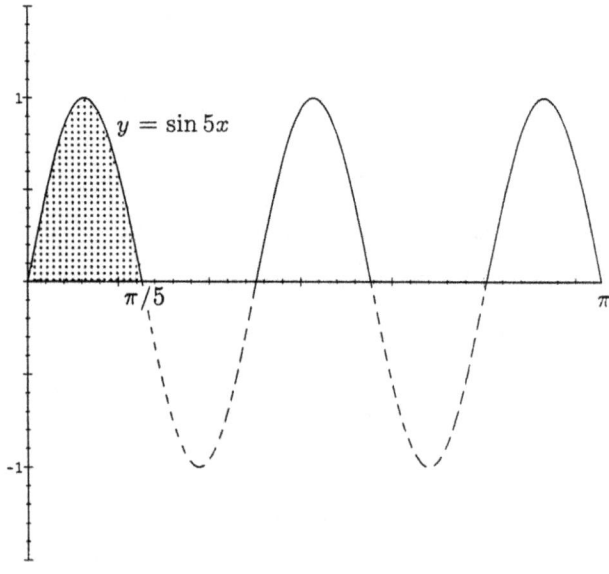

Figure 16.1. The area under one arch of $y = \sin 5x$ is $2/5$.

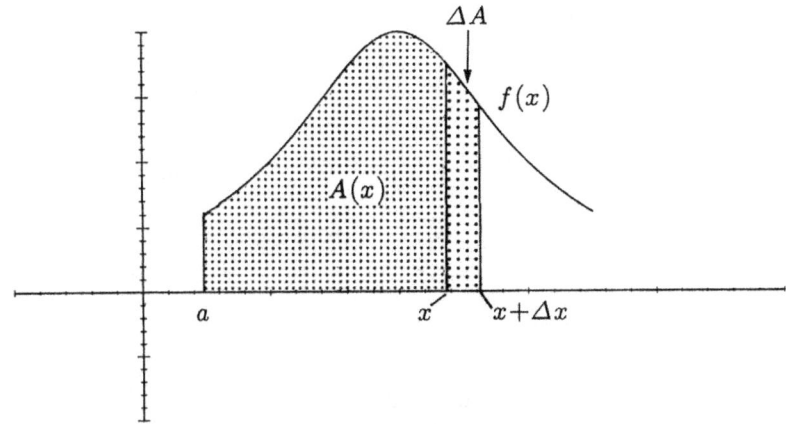

Figure 16.2. The area $A(x)$ under the curve from a to x.

Proof Sketch for Fundamental Theorem

Let $A(x)$ be the area under the curve from a to x:

$$A(x) = \int_a^x f(x)\,dx.$$

The key idea is that the rate of accumulation of area $A'(x)$ as you move to the right is given by how high the graph $y = f(x)$ is:

$$A'(x) = f(x).$$

If $f(x)$ is large, $A(x)$ grows rapidly; if $f(x)$ is small, $A(x)$ grows slowly. More precisely, if x increases a small amount Δx, the corresponding change in area ΔA is approximated by a rectangle of area $=$ (height)(base) $= f(x)\Delta x$:

$$\Delta A \approx f(x)\Delta x; \quad \frac{\Delta A}{\Delta x} \approx f(x).$$

In the limit this approximation becomes exact: $A'(x) = f(x)$. The key idea is established.

The rest of the proof is not too hard. Since $A(x)$ and $F(x)$ have the same derivative, $A(x) = F(x) + C$. To evaluate C, we need to know $A(x)$ for some value of x. But we know that for the starting value $x = a$, $A(a) = 0$:

$$0 = A(a) = F(a) + C \Longrightarrow C = -F(a).$$

Therefore $A(x) = F(x) - F(a)$. In particular, the whole area

$$\int_a^b f(x)\,dx = A(b) = F(b) - F(a).$$

Notation. The antiderivative of $f(x)$ is so closely related to the integral that it is indicated by an integral sign without limits $\int f(x)\,dx$ and called the indefinite integral. For example, $\int \cos x\,dx = \sin x + C$. In contrast, the Riemann integral $\int_a^b f(x)\,dx$ is called the *definite integral*. Note that the indefinite integral is a function, while the definite integral is a number. In this notation, the Fundamental Theorem becomes

$$\int_a^b f(x)\,dx = \int f(x)\,dx\bigg]_a^b.$$

Exercises 16

Evaluate the following definite integrals by using the Fundamental Theorem of Calculus $\int_a^b f(x)\,dx = \int f(x)\,dx\big]_a^b$.

1. $\displaystyle\int_0^1 x\,dx$

2. $\displaystyle\int_0^1 x^2\,dx$

3. $\displaystyle\int_2^4 x\,dx$

4. $\displaystyle\int_2^4 x^2\,dx$

5. $\displaystyle\int_0^{10} x^9 \, dx$

6. $\displaystyle\int_1^{100} \frac{dx}{x^2}$

7. $\displaystyle\int_4^9 \sqrt{x} \, dx$

8. $\displaystyle\int_1^3 \sqrt{x} \, dx$

9. $\displaystyle\int_4^{16} x\sqrt{x} \, dx$

10. $\displaystyle\int_1^3 x\sqrt{x} \, dx$

11. $\displaystyle\int_2^3 \frac{dx}{\sqrt{x}}$

12. $\displaystyle\int_2^3 \frac{dx}{x}$

13. $\displaystyle\int_3^5 e^{x/2} \, dx$

14. $\displaystyle\int_3^5 e^{-x/2} \, dx$

15. $\displaystyle\int_0^{\frac{\pi}{8}} \cos 2x \, dx$

16. $\displaystyle\int_0^{2\pi} \sin \frac{1}{2}x \, dx$

Find the antiderivative as indicated.

17. $\displaystyle\int \sin x \, dx$

18. $\displaystyle\int 3 \, dx$

19. $\displaystyle\int \sin\left(2x - \frac{\pi}{5}\right) dx$

20. $\displaystyle\int xe^{x^2+1} \, dx$

17

Properties of the Definite Integral

As you would expect, the integral of a sum is the sum of the integrals

$$\int_a^b (f(x) + g(x))\, dx = \int_a^b f(x)\, dx + \int_a^b g(x)\, dx \qquad (17.1)$$

and the integral of a multiple is a multiple of the integral

$$\int_a^b cf(x)\, dx = c \int_a^b f(x)\, dx. \qquad (17.2)$$

For example,

$$
\begin{aligned}
\int_0^{\frac{\pi}{18}} \left(\cos 3x + \frac{1}{3}\sin 3x\right) dx &= \int_0^{\frac{\pi}{18}} (\cos 3x)\, dx + \frac{1}{3}\int_0^{\frac{\pi}{18}} (\sin 3x)\, dx \\
&= \left. \frac{1}{3}\sin 3x \right]_0^{\frac{\pi}{18}} + \frac{1}{3}\left[-\frac{1}{3}\cos 3x \right]_0^{\frac{\pi}{18}}
\end{aligned}
$$

$$= \frac{1}{3}\left(\sin\frac{\pi}{6} - \sin 0\right)$$

$$+ \frac{1}{3}\left(-\frac{1}{3}\cos\frac{\pi}{6} - -\frac{1}{3}\cos 0\right)$$

$$= \frac{1}{3}\cdot\frac{1}{2} - \frac{1}{9}\cdot\frac{\sqrt{3}}{2} + \frac{1}{9}\cdot 1 = \frac{5 - \sqrt{3}}{18}.$$

At least property 17.2 seems obvious from the interpretation as area: if you double a function, you double the area under its graph (see Figure 17.1). These two properties and many others can be derived rigorously right from the definition of the integral as a limit of Riemann sums. As an example, here is a derivation of property 17.1.

Proof of Equation 17.1

$$\int_a^b (f(x) + g(x))\,dx = \lim_{\Delta x \to 0} \sum (f(x)\Delta x + g(x)\Delta x)$$

$$= \lim_{\Delta x \to 0} \left(\sum f(x)\Delta x + \sum g(x)\Delta x\right)$$

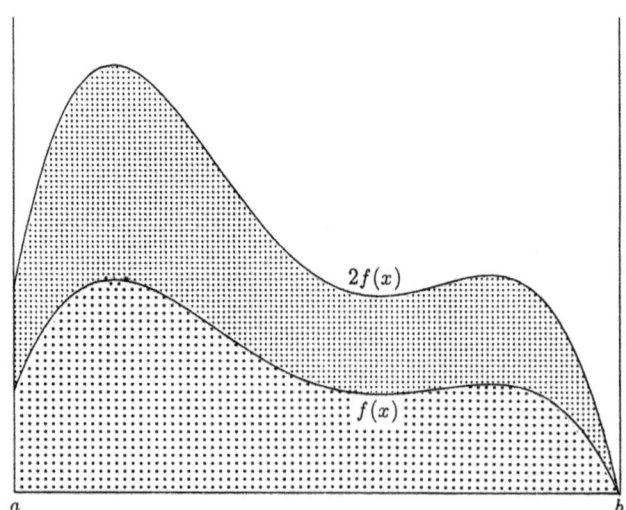

Figure 17.1. Doubling f doubles the area $\int_a^b f(x)\,dx$.

$$= \lim_{\Delta x \to 0} \sum f(x)\Delta x + \lim_{\Delta x \to 0} \sum g(x)\Delta x$$

$$= \int_a^b f(x)\,dx + \int_a^b g(x)\,dx.$$

Not everything you might think might be true is true. For example, the integral of a product of two factors is not generally equal to the product of the integrals. For example,

$$\int_0^1 x^2\,dx = \frac{1}{3}\,x^3\Big]_0^1 = \frac{1}{3},$$

but

$$\left(\int_0^1 x\,dx\right)\left(\int_0^1 x\,dx\right) = \left(\frac{1}{2}\,x^2\Big]_0^1\right)\left(\frac{1}{2}\,x^2\Big]_0^1\right)$$

$$= \frac{1}{2}\cdot\frac{1}{2} = \frac{1}{4}.$$

A third property says that

$$\int_a^c f(x)\,dx = \int_a^b f(x)\,dx + \int_b^c f(x)\,dx. \qquad (17.3)$$

See Figure 17.2.

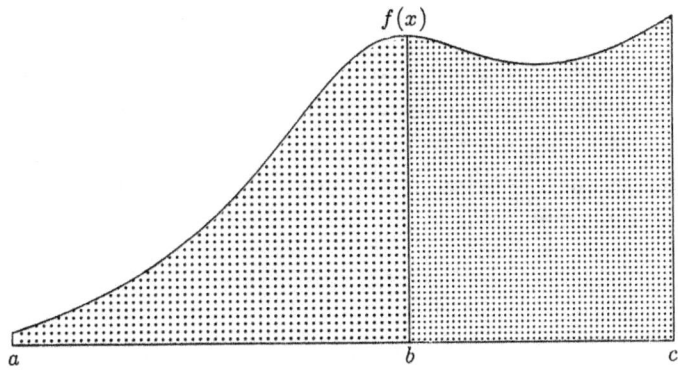

Figure 17.2. The whole area $\int_a^c f(x)\,dx = \int_a^b f(x)\,dx + \int_b^c f(x)\,dx.$

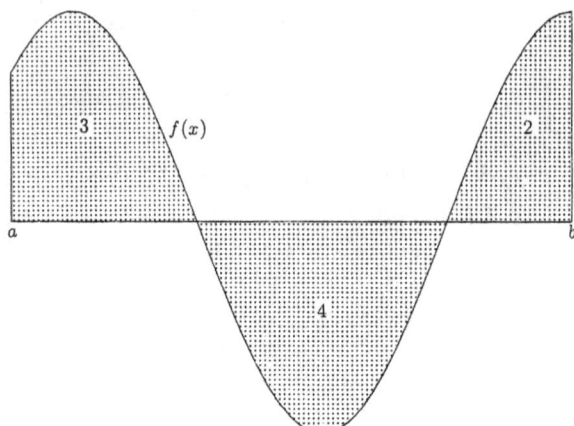

Figure 17.3. If $f(x)$ is sometimes positive and sometimes negative, $\int_a^b f(x)\,dx$ equals the area above the x-axis minus the area below. Here $\int_a^b f(x)\,dx = 3 - 4 + 2 = 1$.

What if $c < a$? If we agree $\int_a^c f(x)\,dx = -\int_c^a f(x)\,dx$, all properties continue to hold.

If $f(x)$ is negative, the area lies under the x-axis and $\int_a^b f(x)\,dx$ is negative. If $f(x)$ is sometimes positive and sometimes negative, $\int_a^b f(x)\,dx$ equals the area above the x-axis minus the area below. See Figure 17.3.

The Area Between Two Curves

To compute the area between two graphs, $f(x) \ge g(x)$, for $a \le x \le b$, cut the interval from a to b into small subintervals of length Δx, as in Figure 17.4. The corresponding strip of area has height about $f(x) - g(x)$ and area about $(f(x) - g(x))\Delta x$. The approximate area is given by the Riemann sum

$$\sum (f(x) - g(x))\Delta x.$$

In the limit, the exact area is given by the integral

$$A = \int_a^b (f(x) - g(x))\,dx.$$

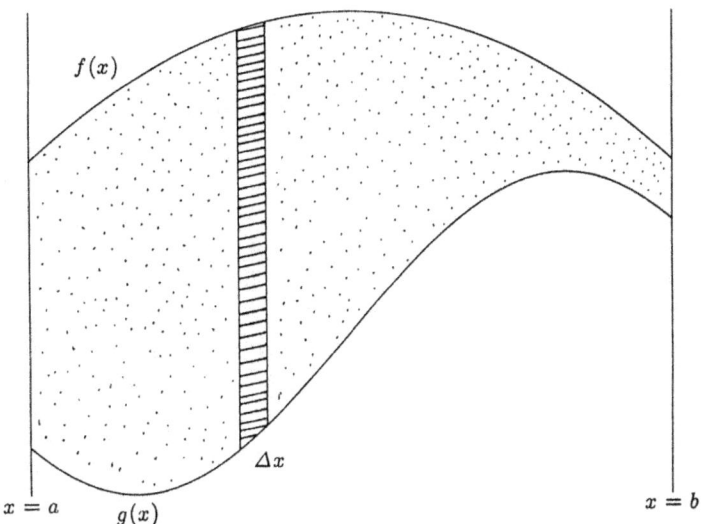

Figure 17.4. The area between two graphs equals $\int_a^b (f(x) - g(x))\, dx$.

This carries the interpretation of slicing the area into infinitely many infinitesimal strips of infinitesimal width dx, height $f(x) - g(x)$, and adding them all up by one integral. Although defined as a limit of Riemann sums, it can often be computed according to the Fundamental Theorem of Calculus by antidifferentiation.

Example 17.1. Compute a single piece of area between the graphs of $\sin x$ and $\cos x$.

We will compute area A of Figure 17.5. All others are the same.

$$
\begin{aligned}
A &= \int_{\frac{\pi}{4}}^{\frac{5\pi}{4}} (\sin x - \cos x)\, dx \\[2mm]
&= \left. -\cos x - \sin x \right]_{\frac{\pi}{4}}^{\frac{5\pi}{4}} \\[2mm]
&= \left(\frac{\sqrt{2}}{2} + \frac{\sqrt{2}}{2} \right) - \left(-\frac{\sqrt{2}}{2} - \frac{\sqrt{2}}{2} \right) = 2\sqrt{2}.
\end{aligned}
$$

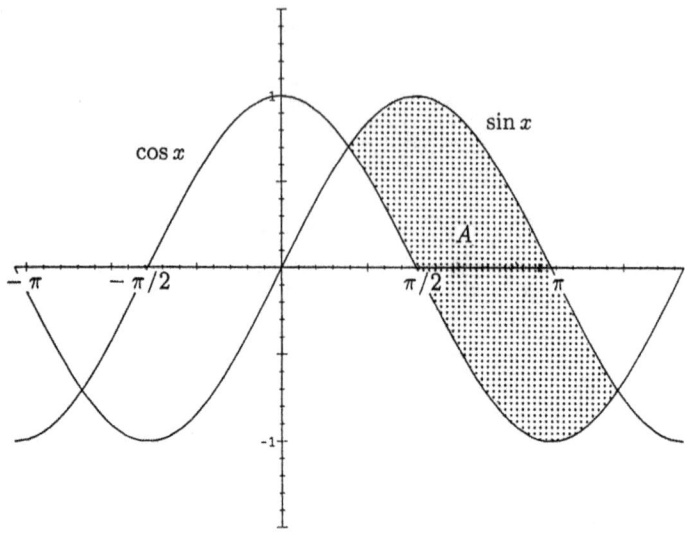

Figure 17.5. A piece of area between the graphs of $\sin x$ and $\cos x$.

Exercises 17

Find the area between $f(x)$ and $g(x)$ for $a \leq x \leq b$. Be sure to integrate the larger one minus the smaller one. Warning: even the algebra can be hard. It is OK to use a calculator to do the arithmetic.

1. $f(x) = x^2 + 2x + 1$, $g(x) = -x^2$, $a = 0$, $b = 2$

2. $f(x) = x^3$, $g(x) = x$, $a = -4$, $b = -2$

3. $f(x) = \dfrac{1}{x^2}$, $g(x) = \dfrac{1}{x^4}$, $a = 1$, $b = 10$

4. $f(x) = \dfrac{1}{x^2}$, $g(x) = \dfrac{1}{x^4}$, $a = 10$, $b = 100$

5. $f(x) = \sqrt{x}$, $g(x) = x^2$, $a = 4$, $b = 16$

6. $f(x) = x\sqrt{x}$, $g(x) = x^3\sqrt{x}$, $a = 1$, $b = 4$

7. $f(x) = \dfrac{1}{x}$, $g(x) = \dfrac{1}{\sqrt{x}}$, $a = 1$, $b = 4$

8. $f(x) = e^{-x}$, $g(x) = \dfrac{1}{x}$, $a = 1$, $b = 4$

9. $f(x) = \cos x$, $g(x) = \sin x$, $a = 0$, $b = \dfrac{\pi}{2}$

10. $f(x) = \cos x$, $g(x) = \sin x$, $a = 0$, $b = \pi$

11. $f(x) = \cos 3x$, $g(x) = \sin 3x$, $a = 0$, $b = \dfrac{\pi}{2}$

Compute.

12. $\displaystyle\int_{2}^{0} (3x^3 - 2x^2)\, dx$

13. $\displaystyle\int_{5}^{1} (e^{5x} - e^{\frac{x}{5}})\, dx$

14. $\displaystyle\int_{a}^{b} (2f(x) + 3g(x))\, dx$, given that

$\displaystyle\int_{a}^{b} f(x)\, dx = \pi$ and $\displaystyle\int_{a}^{b} g(x)\, dx = 7.$

15. $\displaystyle\int_{a}^{b} (f(x) - g(x) + \dfrac{1}{2}h(x))\, dx$, given that

$\displaystyle\int_{a}^{b} f(x)\, dx = 3$, $\displaystyle\int_{a}^{b} g(x)\, dx = 5$, $\displaystyle\int_{a}^{b} h(x)\, dx = 7.$

16. $\displaystyle\int_{3}^{5} f(x)\, dx$, given that

$\displaystyle\int_{3}^{7} f(x)\, dx = 5$ and $\displaystyle\int_{5}^{7} f(x)\, dx = 8$

17. $\displaystyle\int_3^5 f(x)\,dx$ given that

$$\int_3^7 f(x)\,dx = 4,\ \int_5^7 f(x)\,dx\ = 8.$$

18. $\displaystyle\int_5^3 f(x)\,dx$, given that

$$\int_3^7 f(x)\,dx = 2,\ \int_8^5 f(x)\,dx = 11,\ \int_8^7 f(x)\,dx = -100.$$

18

Recognition-Substitution
(For Indefinite Integrals)

Since the derivative of $e^{x^2} + C$ is $e^{x^2}(2x)$, the antiderivative

$$\int e^{x^2} \, 2x \, dx = e^{x^2} + C.$$

When you differentiate, the Chain Rule throws in that $2x$, which you need to antidifferentiate $e^{x^2} \, 2x$ back to $e^{x^2} + C$. To integrate any e^u, you need the du/dx:

$$\int e^u \frac{du}{dx} dx = e^u + C.$$

Comparing that formula with the familiar-looking

$$\int e^u du = e^u + C, \qquad\qquad (18.1)$$

we see that

$$du = \frac{du}{dx} dx,$$

as you might guess. The main point in Equation 18.1 is that to antidifferentiate e^u, you need the

$$du = \frac{du}{dx} dx$$

with the derivative of u there. If you find the e^u and the du both, then the antiderivative is $e^u + C$.

Examples of Equation 18.1

$$\int e^{x^3} 3x^2 dx = \int e^u du = e^u + C = e^{x^3} + C,$$

with $u = x^3$, $du = 3x^2\, dx$.

$$\int x^2 e^{x^3} dx = \frac{1}{3} \int e^{x^3} 3x^2 dx = \frac{1}{3} \int e^u du$$

$$= \frac{1}{3} e^u + C = \frac{1}{3} e^{x^3} + C.$$

$$\int e^{x^3} dx = \;?$$

We cannot integrate the last example, because if $u = x^3$, the $du = 3x^2 dx$ is missing.

General Integral Formulas

The basic antidifferentiation formulas all have that $du = \dfrac{du}{dx} dx$ in them:

$$\int u^r du = \frac{u^{r+1}}{r+1} + C \quad (r \neq -1). \tag{18.2}$$

$$\int \frac{du}{u} = \ln|u| + C. \tag{18.3}$$

$$\int e^u du = e^u + C. \tag{18.4}$$

$$\int \cos u\, du \;=\; \sin u + C. \tag{18.5}$$

$$\int \sin u\, du \;=\; -\cos u + C. \tag{18.6}$$

The General Method of Recognition-Substitution:

A. Pick the integral formula that most closely resembles the worst part of the expression you want to integrate. Write it down.

B. See if it fits by writing equations for u and then du.

Examples 18.1. $\int (x^4+7)^3(4x^3 dx) \quad \int (x^4+7)^3 x^3 dx \quad \int (x^4+7)^3 dx$

For all three of these examples, the two steps of the method are the same:

A. Pick formula (18.2) with $r = 3$ and write

$$\int u^3 du = \frac{u^4}{4} + C.$$

B. Write out the equations for u and du:

$$u = x^4 + 7, \; du = 4x^3 dx.$$

The first example fits perfectly, and therefore

$$\int (x^4 + 7)^3(4x^3 dx) = \frac{(x^4 + 7)^4}{4} + C.$$

The second example is off by a constant, which can be easily adjusted:

$$\int (x^4 + 7)^3 x^3 dx = \frac{1}{4}\int (x^4 + 7)^3(4x^3 dx) = \frac{1}{4}\frac{(x^4 + 7)^4}{4} + C = \frac{1}{16}(x^4 + 7)^4 + C.$$

The third example is missing the du. Of course you cannot adjust by moving variables in and out of the integral like constants. (If you

could, you could do any integral by just moving everything outside!)
This example can be integrated only by expanding out the $(x^4 + 7)^3$
and then antidifferentiating one term at a time:

$$\int (x^4 + 7)^3 dx = \int x^{12} + 21x^8 + 147x^4 + 343 \, dx$$

$$= \frac{1}{13}x^{13} + \frac{7}{3}x^9 + \frac{147}{5}x^5 + 343x + C$$

Example 18.2. $\int \frac{x^2}{\sqrt{x^3+8}} dx$.

A. You could try $\int \frac{du}{u} = ln|u| + C$ or $\int u^{-1/2} du = 2u^{1/2} + C$. The
second is more specific and more likely to work. We'll try both.

B. For $\int \frac{du}{u} = ln|u| + C, u = (x^3+8)^{1/2}, du = \frac{1}{2}(x^3+8)^{-1/2}(3x^2)dx$,
which is much more complicated than the $x^2 dx$ we have; this
formula does not work. For

$$\int u^{-1/2} du = 2u^{1/2} + C,$$

$u = x^3 + 8, du = 3x^2 dx$, and we can adjust:

$$\int \frac{x^2}{\sqrt{x^3+8}} dx = \frac{1}{3}\int \frac{3x^2}{\sqrt{x^3+8}} dx = \frac{1}{3}2(x^3+8)^{1/2} + C = \frac{2}{3}\sqrt{x^3+8} + C.$$

Acknowledgment. This chapter emerged from the questions and
suggestions of Math 103 Fall 1991 and Math 104 Fall 1999, Williams
College. Recognition is a simple kind of substitution, the topic of
the next chapter.

Exercises 18

1. $\int (5x + 3)^{99} dx$

2. $\int \sqrt{7x - 5} \, dx$

3. $\displaystyle\int e^{4x-1}\,dx$

4. $\displaystyle\int \cos\left(4x - \frac{\pi}{12}\right)\,dx$

5. $\displaystyle\int \sin(3x + \pi)\,dx$

6. $\displaystyle\int \frac{dx}{4x - 3}$

7. $\displaystyle\int \left(5x^2 + 3\right)^{99} x\,dx$

8. $\displaystyle\int xe^{4x^2}\,dx$

9. $\displaystyle\int x^3 \cos(4x^4 - 3)\,dx$

10. $\displaystyle\int \frac{x\,dx}{1 + x^2}$

11. $\displaystyle\int x\sqrt{x^2 + 1}\,dx$

12. $\displaystyle\int \frac{x}{\sqrt{x^2 + 1}}\,dx$

13. $\displaystyle\int \sin^2 x \cos x\,dx$

14. $\displaystyle\int \sin^2 3x \cos 3x\,dx$

15. $\displaystyle\int_{x=0}^{10} xe^{-x^2}\,dx$

16. $\displaystyle\int_{x=0}^{\pi/4} \cos^3 x \sin x \, dx$

17. $\displaystyle\int_{x=0}^{\sqrt[3]{26}} \frac{x^2 \, dx}{x^3 + 1}$

18. $\displaystyle\int_{x=0}^{2} x\sqrt{2x^2 + 1} \, dx$

19. $\displaystyle\int_{x=0}^{2} \frac{x \, dx}{\sqrt{2x^2 + 1}}$

*20. $\displaystyle\int_{x=0}^{\infty} e^{-x^2} \, dx$

21. $\displaystyle\int \frac{x \, dx}{\sqrt[3]{5 - x^2}}$

22. $\displaystyle\int \frac{x^2 \, dx}{\sqrt[5]{x^3 + 9}}$

23. $\displaystyle\int \frac{x \, dx}{(x^2 + 7)^2}$

24. $\displaystyle\int \frac{x^2 \, dx}{(x^3 + 7)^3}$

25. $\displaystyle\int x(x^2 + 3)^5 \, dx$

26. $\displaystyle\int (x^2 + 3)^5 \, dx$

19

Integration by Substitution

Many integrals that do not quite look like ones we can antidifferentiate can be adjusted by substituting another variable for x. Let's start with an old example:

$$\int_{x=0}^{3} xe^{x^2}\, dx = \frac{1}{2}\int_{x=0}^{3} e^{x^2}(2x\,dx)$$

$$= \frac{1}{2}e^{x^2}\Big]_{0}^{3} = \frac{1}{2}(e^9 - 1).$$

Even if you did not recognize the integral above as $\frac{1}{2}\int e^u du$, you could still try the substitution $x^2 = u$. Then $x = u^{\frac{1}{2}}$, $\dfrac{dx}{du} = \dfrac{1}{2}u^{-\frac{1}{2}}$, so $dx = \dfrac{1}{2}u^{-\frac{1}{2}}du$. Consequently

$$\int xe^{x^2} dx = \int u^{\frac{1}{2}}e^u \frac{1}{2}u^{-\frac{1}{2}}\, du$$

$$= \frac{1}{2}\int e^u du = \frac{1}{2}e^u + C = \frac{1}{2}e^{x^2} + C.$$

Table 19.1. Common Substitutions

If you see	Try the substitution
$ax + b$	$u = ax + b$
x^p	$u = x^p$
$a^2 - x^2$ (esp. $\sqrt{a^2 - x^2}$)	$x = a \sin \theta$
$a^2 + x^2$	$x = a \tan \theta$

Thus

$$\int_{x=0}^{3} xe^{x^2}\, dx = \frac{1}{2}e^{x^2}\bigg]_0^3 = \frac{1}{2}(e^9 - 1).$$

For a more complicated definite integral, you can avoid changing the integral back from u to x by changing the limits of integration to u. Here since $u = x^2$, when $x = 0$, $u = 0$, and when $x = 3$, $u = 9$. Therefore,

$$\int_{x=0}^{3} xe^{x^2}\, dx = \int_{u=0}^{9} \frac{1}{2}e^u\, du$$

$$= \frac{1}{2}e^u\bigg]_{u=0}^{9} = \frac{1}{2}(e^9 - 1).$$

To remember whether it's x or u, I like to write '$x = 0$' instead of just '0' as the lower limit.

Sometimes it is hard to know what substitution to try; sometimes you try a lot and none work. Table 19.1 gives some suggestions. Note for example that the substitution $x = a \sin \theta$ turns $\sqrt{a^2 - x^2}$ into

$$\sqrt{a^2 - a^2\sin^2\theta} = a\sqrt{1 - \sin^2\theta} = a \cos \theta,$$

a much simpler expression (assuming a, $\cos \theta > 0$).

Example 19.1. Find

$$\int_{x=0}^{\frac{3}{2}} \frac{dx}{\sqrt{9 - x^2}}.$$

Solution. Since you see $\sqrt{9-x^2} = \sqrt{a^2-x^2}$ with $a = 3$, try the substitution $x = a\sin\theta = 3\sin\theta$. First change the limits of integration: when $x = 0$, $3\sin\theta = 0$, $\theta = 0$. When $x = 3/2$, $3\sin\theta = 3/2$, $\sin\theta = 1/2$, $\theta = \pi/6$.

$$\frac{1}{\sqrt{9-x^2}} = \frac{1}{\sqrt{9-(3\sin\theta)^2}}$$

$$= \frac{1}{\sqrt{9-9\sin^2\theta}} = \frac{1}{\sqrt{9}\sqrt{1-\sin^2\theta}} = \frac{1}{3\cos\theta}$$

(with a plus sign since when $0 \le \theta \le \pi/6$, $\cos\theta \ge 0$). Since $x = 3\sin\theta$, $dx = 3\cos\theta\, d\theta$. The integral becomes

$$\int_{\theta=0}^{\pi/6} \frac{3\cos\theta\, d\theta}{3\cos\theta} = \int_{\theta=0}^{\pi/6} 1\, d\theta = \theta]_{\theta=0}^{\pi/6} = \pi/6.$$

Example 19.2. Find the area of a circle of radius a.

Solution. On the circle $x^2 + y^2 = a^2$, $y = \pm\sqrt{a^2-x^2}$, with the plus sign on the top semicircle where y is positive and the minus sign on the bottom semicircle where y is negative. $-a \le x \le a$. See Figure 19.1.

The area of the circle is the area between the two curves:

$$A = \int_{-a}^{a} (\sqrt{a^2-x^2} - -\sqrt{a^2-x^2})\, dx = 2\int_{x=-a}^{a} \sqrt{a^2-x^2}\, dx.$$

Try $x = a\sin\theta$. Remember to change the limits of integration: when $x = -a$, $\sin\theta = -1$, $\theta = -\pi/2$; when $x = a$, $\sin\theta = 1$, $\theta = \pi/2$.

$$\sqrt{a^2-x^2} = \sqrt{a^2-a^2\sin^2\theta} = \sqrt{a^2}\sqrt{1-\sin^2\theta} = a\cos\theta$$

(with a plus sign since when $-\pi/2 \le \theta \le \pi/2$, $\cos\theta \ge 0$); dx becomes $a\cos\theta\, d\theta$, and the integral becomes

$$A = 2\int_{\theta=-\pi/2}^{\pi/2} (a\cos\theta)(a\cos\theta\, d\theta) = 2a^2 \int_{\theta=-\pi/2}^{\pi/2} \cos^2\theta\, d\theta.$$

You may think it looks hopeless to integrate $\cos^2 x$, but it is one of the most frequently used tricks in the business. Just memorize and

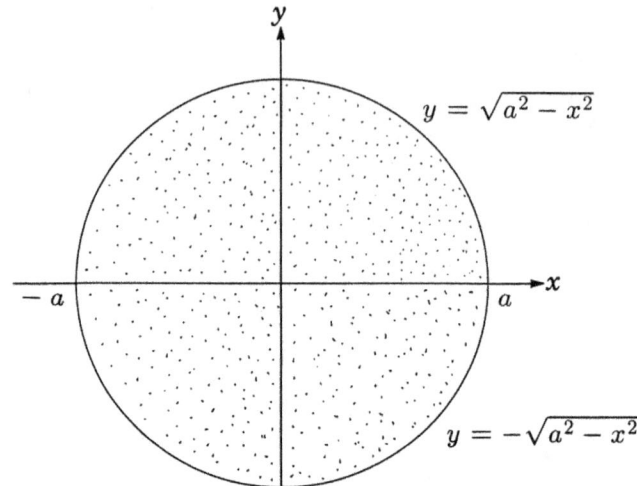

Figure 19.1. The area of the circle is the area between the two curves $y = +\sqrt{a^2 - x^2}$ and $y = -\sqrt{a^2 - x^2}$.

use the identity (see Table 21.2):

$$\cos^2 \theta = \frac{1}{2} + \frac{1}{2} \cos 2\theta.$$

Then

$$
\begin{aligned}
A &= 2a^2 \int_{\theta=-\pi/2}^{\pi/2} \left(\frac{1}{2} + \frac{1}{2} \cos 2\theta \right) d\theta \\
&= 2a^2 \left[\frac{1}{2}\theta + \frac{1}{4} \sin 2\theta \right]_{-\pi/2}^{\pi/2} \\
&= 2a^2 \left(\frac{\pi}{4} - \left(-\frac{\pi}{4} \right) \right) = \pi a^2,
\end{aligned}
$$

the area of a circular disc.

In multivariable calculus, you will see a much easier computation of the area of a circle using polar coordinates.

Exercises 19

1. $\displaystyle\int_{x=0}^{1} \frac{dx}{\sqrt{4-x^2}}$

2. $\displaystyle\int_{x=0}^{\frac{3}{2}} \frac{dx}{\sqrt{3-x^2}}$

3. $\displaystyle\int_{x=0}^{\frac{3}{2}} \sqrt{9-x^2}\, dx$

4. $\displaystyle\int_{x=0}^{\frac{1}{2\sqrt{3}}} \sqrt{\frac{1}{9}-x^2}\, dx$

5. $\displaystyle\int_{x=0}^{\frac{3}{2}} \frac{x\, dx}{\sqrt{3-x^2}}$ (Do 2 ways: as $\displaystyle\int u^{-1/2} du$ and by $x = \sqrt{3}\sin\theta$.)

6. $\displaystyle\int_{x=0}^{\frac{3}{2}} x\sqrt{9-x^2}\, dx$ (Do 2 ways.)

7. $\displaystyle\int_{x=0}^{\sqrt{3}} x\sqrt{3-x^2}\, dx$ (Do 2 ways.)

*8. $\displaystyle\int_{x=-1}^{1} \sqrt{3-2x-x^2}\, dx$ (Hint: integrand is $\sqrt{4-(x+1)^2}$)

9. $\displaystyle\int \frac{x\, dx}{7x+3}$

10. $\displaystyle\int_{x=0}^{\frac{1}{4}} \sqrt{1-4x^2}\, dx$

11. $\displaystyle\int_{x=0}^{\frac{1}{\sqrt{2}}} \sqrt{2-3x^2}\, dx$

20

Review of Integration

$$\int u^r \, du = \frac{u^{r+1}}{r+1} + C \quad (r \neq -1)$$

$$\int \frac{du}{u} = \ln |u| + C$$

$$\int e^u \, du = e^u + C \qquad\qquad \int e^{ax} \, dx = \frac{1}{a} e^{ax} + C$$

$$\int \cos u \, du = \sin u + C \qquad \int \cos ax \, dx = \frac{1}{a} \sin ax + C$$

$$\int \sin u \, du = -\cos u + C \qquad \int \sin ax \, dx = -\frac{1}{a} \cos ax + C$$

$$\int \cos^2 \theta \, d\theta = \int \left(\frac{1}{2} + \frac{1}{2} \cos 2\theta \right) d\theta = \frac{\theta}{2} + \frac{1}{4} \sin 2\theta + C$$

$$\int \sin^2 \theta \, d\theta = \int \left(\frac{1}{2} - \frac{1}{2} \cos 2\theta \right) d\theta = \frac{\theta}{2} - \frac{1}{4} \sin 2\theta + C$$

If you see $a^2 - x^2$ (esp. $\sqrt{a^2 - x^2}$), try substituting $x = a\sin\theta$, $dx = a\cos\theta d\theta$, $\sqrt{a^2 - x^2} = a\cos\theta$.

Definition of Definite Integral: $\displaystyle\int_a^b f(x)dx = \lim_{\Delta x \to 0} \sum f(x)\Delta x.$

Fundamental Theorem of Calculus: $\displaystyle\int_a^b f(x)\,dx = F(b) - F(a) =$

$F(x)]_a^b$, where $F(x)$ is any antiderivative of $f(x)$.

21

Trigonometric Functions and Their Inverses

Once you understand sine and cosine, you can deduce the behavior of the other trigonometric functions. For example

$$\tan x = \frac{\sin x}{\cos x}.$$

It is 0 when $\sin x$ is 0, namely, when x is a multiple of π. It blows up when $\cos x$ is 0, namely, when

$$x = \ldots - \frac{3\pi}{2}, -\frac{\pi}{2}, \frac{\pi}{2}, \frac{3\pi}{2}, \ldots.$$

Its derivative can be computed by the Quotient Rule as

$$(\tan x)' = \frac{\cos^2 x + \sin^2 x}{\cos^2 x} = \frac{1}{\cos^2 x} = \sec^2 x,$$

by the identity $\cos^2 x + \sin^2 x = 1$, and the definition

$$\sec x = \frac{1}{\cos x}.$$

Table 21.1. The trigonometric functions and their derivatives.

Function	Derivative
$\sin x$	$\cos x$
$\cos x$	$-\sin x$
$\tan x = \dfrac{\sin x}{\cos x}$	$\sec^2 x$
$\sec x = \dfrac{1}{\cos x}$	$\sec x \tan x$
$\cot x = \dfrac{\cos x}{\sin x}$	$-\csc^2 x$
$\csc x = \dfrac{1}{\sin x}$	$-\csc x \cot x$

The definitions and derivatives of the trigonometric functions appear in Table 21.1. Tangent and secant are the important ones to memorize at once.

Example 21.1. If $y = \tan x^3$, find y'.

Solution. $y' = (\sec^2 x^3)(x^3)' = (\sec^2 x^3)(3x^2) = 3x^2 \sec^2 x^3$.

Table 21.2 gives some trigonometric identities: right now is the time to get the first four by heart.

21.1. Inverse Trigonometric Functions

Just as we needed the inverse of the exponential function $y = e^x$, namely $x = \ln y$, we will need inverses of the trigonometric functions. The inverse sine function, just written $\sin^{-1} x$, means "the angle whose sine is x." For example, $\sin^{-1} 1/2 = \pi/6$, because $\sin \pi/6 = 1/2$. Actually there are other angles θ with $\sin \theta = 1/2$, such as $\theta = \pi/6 + 2\pi$, but we agree to choose ones between $-\pi/2$ and

Table 21.2. Trigonometric identities.

$$\sin^2 x + \cos^2 x = 1$$

$$\tan^2 x + 1 = \sec^2 x$$

$$\cos^2 x = \tfrac{1}{2} + \tfrac{1}{2}\cos 2x$$

$$\sin^2 x = \tfrac{1}{2} - \tfrac{1}{2}\cos 2x$$

$$\cos(A+B) = \cos A \cos B - \sin A \sin B$$

$$\cos 2x = \cos^2 x - \sin^2 x$$

$$\sin(A+B) = \sin A \cos B + \cos A \sin B$$

$$\sin 2x = 2\sin x \cos x$$

$\pi/2$, which cover all possible values of sine from -1 to 1. Similarly we choose $\cos^{-1} x$ between 0 and π, which cover all possible values of cosine from 1 to -1, and $\tan^{-1} x$ between $-\pi/2$ and $\pi/2$, which cover all possible values of tangent between $-\infty$ and $+\infty$. For example, $\cos^{-1}(-1/2) = 2\pi/3$, $\tan^{-1}(-1) = -\pi/4$.

This notation is different from $\sin^2 x = (\sin x)^2$, since $\sin^{-1} x \neq (\sin x)^{-1}$. We agree that $\sin^p x = (\sin x)^p$ except for $p = -1$, when $\sin^{-1} x$ denotes the inverse function. Sometimes $\sin^{-1} x$ is called arcsin x.

Example 21.2. Find $\cos^{-1}(-1)$.

Solution. $\cos^{-1}(-1)$ means the angle whose cosine is -1, namely, π, because $\cos \pi = -1$. Therefore $\cos^{-1}(-1) = \pi$.

Example 21.3. Simplify the expression $\cos \sin^{-1} x$.

Solution. The expression $\cos \sin^{-1} x$ means the cosine of the angle θ whose sine is x. In other words, if $\sin \theta = x$, what is $\cos \theta$? The best general method is to draw a triangle, as in Figure 21.1.

Label x and 1 to make $\sin \theta = x/1$. Use Pythagoras to label the third side $\sqrt{1 - x^2}$.

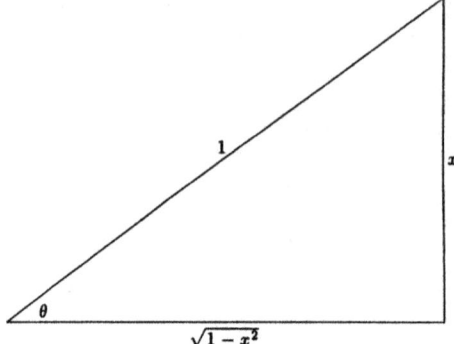

Figure 21.1. This triangle shows how to compute that $\cos\sin^{-1} x = \sqrt{1-x^2}$.

Read off

$$\cos\theta = \sqrt{1-x^2}/1 = \sqrt{1-x^2}.$$

Therefore $\cos\sin^{-1} x = \sqrt{1-x^2}$.

To find the derivative of $y = \sin^{-1} x$, write $x = \sin y$ and differentiate implicitly: $1 = (\cos y)y'$, so that $y' = 1/\cos y$. In terms of x, $y' = 1/\cos\sin^{-1} x = 1/\sqrt{1-x^2}$. Table 21.3 gives the three most important inverse trigonometric functions and their derivatives.

Example 21.4. If $y = \tan^{-1} x^3$, find y'.

Solution. $y' = \dfrac{1}{1+(x^3)^2}(x^3)' = \dfrac{3x^2}{1+x^6}$.

Table 21.3. Inverse trigonometric functions and derivatives.

Function	Range	Derivative
$y = \sin^{-1} x$	$-\frac{\pi}{2} \leq y \leq \frac{\pi}{2}$	$y' = \dfrac{1}{\sqrt{1-x^2}}$
$y = \tan^{-1} x$	$-\frac{\pi}{2} < y < \frac{\pi}{2}$	$y' = \dfrac{1}{1+x^2}$
$y = \cos^{-1} x$	$0 \leq y \leq \pi$	$y' = -\dfrac{1}{\sqrt{1-x^2}}$

Table 21.4. Inverse trigonometric functions as antiderivatives.

Function $f(x)$	Antiderivative $F(x)$
$\dfrac{1}{\sqrt{a^2 - x^2}}$	$\sin^{-1}\dfrac{x}{a} + C$
$\dfrac{1}{a^2 + x^2}$	$\dfrac{1}{a}\tan^{-1}\dfrac{x}{a} + C$

As a consequence, we know a couple of important new antiderivatives, listed in Table 21.4, which can be verified by differentiating. For example,

$$\left(\sin^{-1}\frac{x}{a}\right)' = \frac{1}{\sqrt{1 - \frac{x^2}{a^2}}} \cdot \frac{1}{a} = \frac{1}{\sqrt{a^2 - x^2}}.$$

Combining with the Chain Rule yields

$$\int \frac{du}{\sqrt{a^2 - u^2}} = \sin^{-1}\frac{u}{a} + C. \qquad (21.1)$$

$$\int \frac{du}{a^2 + u^2} = \frac{1}{a}\tan^{-1}\frac{u}{a} + C. \qquad (21.2)$$

Examples 21.5.

$$\int \frac{dx}{5 + x^2} = \frac{1}{\sqrt{5}}\tan^{-1}\frac{x}{\sqrt{5}} + C.$$

$$\int \frac{x^2 dx}{1 + (x^3)^2} = \frac{1}{3}\int \frac{3x^2 dx}{1 + (x^3)^2} = \frac{1}{3}\tan^{-1} x^3 + C.$$

(This second example used Equation 21.2 with $u = x^3$.)

Exercises 21

Find the derivatives y'.

1. $y = 3\tan x$

2. $y = \tan 3x$

3. $y = \tan x^3$

4. $y = \tan^3 x$ (This means $(\tan x)^3$.)

5. $y = \tan \sqrt{x}$

6. $y = \sec x^2$

7. $y = \sec(x^2 + 3x - 7)$

8. $y = \ln(\sec x + \tan x)$

9. $y = x \cot x$

10. $y = x^2 \csc^2 x^2$

Simplify.

11. $\dfrac{\sqrt{1 - \sin^2 x}}{\sqrt{1 - \cos^2 x}}$ $(0 < x < \frac{\pi}{2})$

12. $\sin^4 x + 2\sin^2 x \cos^2 x + \cos^4 x$

13. $(\sec^2 x - 1)^2$

14. If $\cos 2x = \frac{1}{3}$, find $\cos^2 x$ and $\sin^2 x$.

15. If $\cos A = \frac{3}{5}$ and $\cos B = \frac{12}{13}$, find $\cos(A + B)$ $(0 < A, B < \frac{\pi}{2})$.

16. $\sin^{-1} \dfrac{\sqrt{3}}{2}$

17. $\sin^{-1}\left(-\dfrac{\sqrt{3}}{2}\right)$

18. $\cos^{-1}\dfrac{\sqrt{3}}{2}$

19. $\cos^{-1}\left(-\dfrac{\sqrt{3}}{2}\right)$

20. $\tan^{-1}0$

21. $\tan^{-1}\sqrt{3}$

22. $\tan^{-1}(-\sqrt{3})$

23. $\cos\sin^{-1}\frac{1}{5}$

24. $\cos\tan^{-1}\frac{1}{5}$

25. $\tan\cos^{-1}\frac{3}{4}$

26. $\sin\cos^{-1}\frac{3}{4}$

Compute the derivative y':

27. $y = 5\sin^{-1}x$

28. $y = -\sin^{-1}5x$

29. $y = \left(\sin^{-1}x\right)^{5}$

30. $y = \sin^{-1}\sqrt{x}$

31. $y = \tan^{-1}3x$

32. $y = \tan^{-1}(x^{2}+1)$

33. $y = \tan^{-1}\sqrt{x}$

34. $y = \cos^{-1}(e^x)$

35. $y = e^{\tan^{-1} x}$

36. $y = 7e^{5 \tan^{-1} 3x}$

37. Derive the formula $\left(\tan^{-1} x\right)' = \dfrac{1}{1 + x^2}$

38. Derive the formula $\left(\cos^{-1} x\right)' = \dfrac{1}{\sqrt{1 - x^2}}$.

39. $\displaystyle\int_{x=0}^{\frac{1}{2}} \dfrac{dx}{\sqrt{1 - x^2}}$ (Do both by formula and by trig substitution.)

40. $\displaystyle\int_{x=0}^{2} \dfrac{dx}{\sqrt{4 - x^2}}$ (Do both by formula and by trig substitution.)

41. $\displaystyle\int_{x=-1}^{1} \dfrac{dx}{1 + x^2}$ (Do both by formula and by trig substitution.)

42. $\displaystyle\int_{0}^{\sqrt{3}} \dfrac{dx}{5 + x^2}$

43. $\displaystyle\int \dfrac{dx}{1 + (2x - 3)^2}$

44. $\displaystyle\int \dfrac{dx}{\sqrt{1 - (4x - 1)^2}}$

45. $\displaystyle\int \dfrac{2x\,dx}{1 + (x^2)^2}$

22

Volume, Length, Average

22.1. Volume

Example 22.1. Find the volume enclosed by revolving one arch of $y = \sin x$ about the x-axis. (See Figure 22.1.)

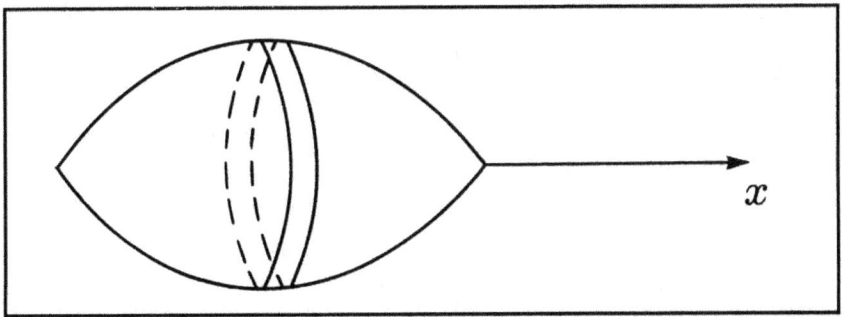

Figure 22.1. This volume of revolution can be computed by slicing it into infinitesimally thin discs and integrating.

Solution. *Always start by guessing the answer.* This volume is less than a sphere's of radius $\pi/2$, which is

$$\frac{4}{3}\pi \left(\frac{\pi}{2}\right)^3 \approx \frac{1}{6}\cdot 3^4 \approx 13,$$

but more than a sphere's of radius 1, which is $\frac{4}{3}\pi \approx 4$. I guess around 8.

Slice the volume up into infinitesimally thin discs of thickness dx, radius $y = \sin x$, and volume $dV = \pi y^2 dx = \pi \sin^2 x dx$. To get the total volume, add up the contributions of each slice by the integral

$$\int_{x=0}^{\pi} \pi \sin^2 x dx.$$

Notice that since we have dx, everything must be in terms of x; we could not have left the y^2 in there. The discs range from tiny ones at the left $(x = 0)$ to tiny ones at the right $(x = \pi)$.

To integrate $\sin^2 x$, just use the identity you have memorized

$$\sin^2 x = \frac{1}{2} - \frac{1}{2}\cos 2x.$$

Then

$$V = \pi \int_{x=0}^{\pi} \left(\frac{1}{2} - \frac{1}{2}\cos 2x\right) dx$$

$$= \pi \left[\frac{1}{2}x - \frac{1}{4}\sin 2x\right]_0^{\pi} = \frac{\pi^2}{2} \approx 5,$$

which is at least within the range of my guess.

22.2. Arclength

By Pythagoras, an infinitesimal element of arclength

$$ds = \sqrt{dx^2 + dy^2} = \sqrt{1 + \left(\frac{dy}{dx}\right)^2}\, dx = \sqrt{\left(\frac{dx}{dt}\right)^2 + \left(\frac{dy}{dt}\right)^2}\, dt.$$

See Figure 22.2. The first formula is easy to remember. The second is good if you are given y as a function of x, and the third is good if you

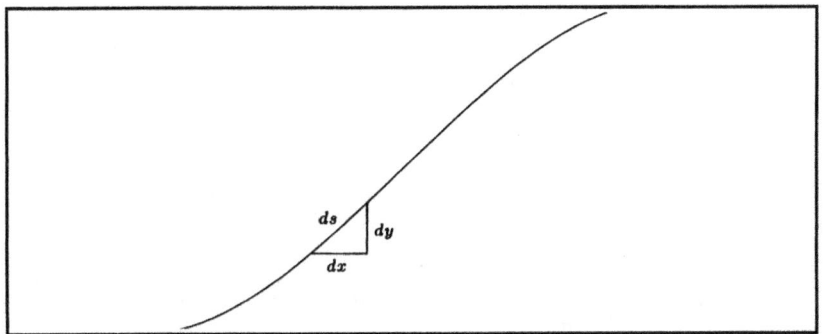

Figure 22.2. By Pythagoras, an infinitesimal element of arclength $ds = \sqrt{dx^2 + dy^2}$.

are given x and y as functions of some more convenient parameter t. The arclength is given by integrating:

$$\text{arclength} = \int \sqrt{1 + (\frac{dy}{dx})^2} \, dx = \int \sqrt{(\frac{dx}{dt})^2 + (\frac{dy}{dt})^2} \, dt.$$

Example 22.2. Find the length of an arc of a circle of radius a through an angle of θ_0. See Figure 22.3. On the circle, $x = a\cos\theta, y = a\sin\theta$, and x and y are given as functions of $t = \theta$.

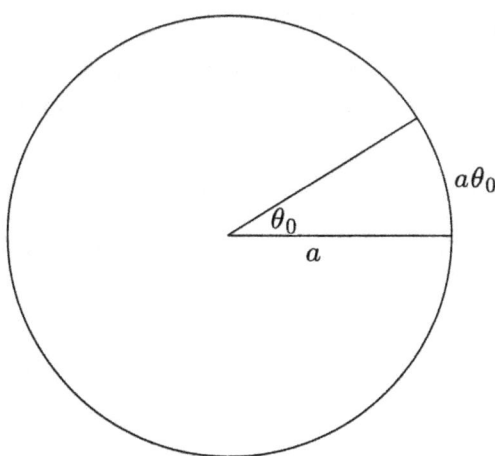

Figure 22.3. The length of an arc of a circle is $a\theta_0$.

$$ds = \sqrt{\left(\frac{dx}{d\theta}\right)^2 + \left(\frac{dy}{d\theta}\right)^2} \, dt$$

$$= a\sqrt{\sin^2\theta + \cos^2\theta} \, d\theta = a \, d\theta.$$

Hence

$$\text{length} = \int_{\theta=0}^{\theta_0} ds = \int_{\theta=0}^{\theta_0} a \, d\theta = a\theta\big]_0^{\theta_0} = a\theta_0.$$

This answer makes good sense: the length should be the fraction $\theta_0/2\pi$ of the circumference $2\pi a$: $(\theta_0/2\pi)(2\pi a) = a\theta_0$.

22.3. Average

The area under a curve ought to be the length times the average height. The value of any integral $\int_a^b f(x)dx$ ought to be $b - a$ times the average value $\bar{f}$ of f. Actually, this idea is used to define the average value $\bar{f}$ of $f(x)$ on an interval $[a, b]$:

$$\bar{f} = \frac{1}{b-a} \int_a^b f(x) \, dx.$$

Example 22.3. Find the average value of $\sin x$ on $[0, \pi]$.

Solution. Since it varies between 0 and 1, you might guess the average is $1/2$. The average value on $[0, \pi]$ is actually

$$\frac{1}{\pi} \int_0^\pi \sin x \, dx = \frac{2}{\pi} \approx .6366$$

($\sin x$ is bigger than $1/2$ more of the time than it is less than $1/2$). In contrast, the average value on $[0, 2\pi]$ is 0.

Exercises 22

1. Find the volume enclosed by revolving the region between $y = 4 - x^2$ and the x-axis around the x-axis.

2. Find the volume enclosed by revolving the region bounded by $y = e^x$, $x = 0$, $x = 3$, $y = 0$ about the x-axis.

3. Find the volume enclosed by revolving the region bounded by the axes and the line $y = 4 - 2x$ about the y-axis.

4. Compute the volume of a sphere of radius a.

5. Find the length of the graph of $y = x^{\frac{3}{2}}$ for $1 \leq x \leq 4$.

6. Find the length of the graph of $y = \frac{1}{3}(x^2 + 2)^{\frac{3}{2}}$ for $1 \leq x \leq 5$.

7. What is the average value of $\cos^2 x$ for $0 \leq x \leq 2\pi$?

8. What is the average value of x^n for $0 \leq x \leq 1$?

9. What is the average value of e^x for $0 \leq x \leq 100$? Is it more or less than $\dfrac{e^0 + e^{100}}{2}$?

10. What is the average value of $\sin^{-1} x$ for $0 \leq x \leq 1$?

23

Integration by Table

For simple integrals, it is much easier and faster just to use the methods you have learned. At the other extreme, many difficult integrals have no antiderivatives in terms of functions you have ever encountered, and you would have to resort to estimates or numerical integration (Chapter 25). But in many intermediate cases an extensive table for formulas will remain useful, at least until computerized symbolic integration is readily available. Some time check out the awesome integral tables in a CRC handbook in the reference section at your library. For now, you can enjoy the ones inside the covers of this book. Unfortunately, the problems that arise do not quite match up with the integral tables.

Example 23.1.

$$\int \cos^3 \left(2\theta - \frac{\pi}{6} \right) d\theta.$$

This is close to Formula 8 from the table, with $u = 2\theta - \frac{\pi}{6}$, $du = 2\,d\theta$. It exactly equals half the table entry:

$$\frac{1}{2}\int \cos^3\left(2\theta - \frac{\pi}{6}\right)(2\,d\theta)$$

$$= \frac{1}{2}\left(\sin\left(2\theta - \frac{\pi}{6}\right) - \frac{1}{3}\sin^3\left(2\theta - \frac{\pi}{6}\right)\right) + \text{C}.$$

Example 23.2.

$$\int \cos^3 x^2\,dx.$$

This is also close to Formula 8 from the table, with $u = x^2$, $du = 2x\,dx$. But that $du = 2x\,dx$ spoils things. Of course one cannot bring out a $\frac{1}{2x}$ in front of the integral; that only works for constants. (Otherwise you could do any integral by bringing the whole integrand function out!) For this integral, one must resort to numerical methods (Chapter 13).

Sometimes you need to use reduction formulas, such as the second version of 17 from the table, repeatedly. Here we take $n = 0$.

$$\int \cos^8\theta\,d\theta = \frac{\cos^7\theta\sin\theta}{8} + \frac{7}{8}\int \cos^6\theta\,d\theta$$

$$= \frac{\cos^7\theta\sin\theta}{8} + \frac{7}{8}\left(\frac{\cos^5\theta\sin\theta}{6} + \frac{5}{6}\int \cos^4\theta\,d\theta\right)$$

$$= \frac{1}{8}\cos^7\theta\sin\theta + \frac{7}{48}\cos^5\theta\sin\theta$$

$$+ \frac{7\cdot 5}{8\cdot 6}\left(\frac{3}{8}\theta + \frac{1}{4}\sin 2\theta + \frac{1}{32}\sin 4\theta\right) + \text{C}$$

by Formula 10 from the table.

Warning. It seems that the only way to get these answers right is to write out everything in excruciating detail.

23.1. Completing the Square

Completing the square is the most important method in using tables. Often what appear to be impossible integrals can be made to match the tables by this method, which reduces any quadratic expression

$$A_0 x^2 + B_0 x + C_0$$

to a multiple of the more familiar $u^2 \pm a^2$. First factor out the A_0 to get something of the form $x^2 + Bx + C$. Then notice that

$$x^2 + Bx + C = \left(x + \frac{1}{2}B\right)^2 + \left(C - \frac{1}{4}B^2\right)$$

$$= u^2 \pm a^2$$

where

$$u = x + \frac{1}{2}B \quad \text{and} \quad a = \sqrt{|C - \frac{1}{4}B^2|}.$$

Example 23.3.

$$\int \frac{dx}{\sqrt{3x^2 + 9x - 4}} = \frac{1}{\sqrt{3}} \int \frac{dx}{\sqrt{x^2 + 3x - \frac{4}{3}}}$$

$$= \frac{1}{\sqrt{3}} \int \frac{dx}{\sqrt{\left(x + \frac{3}{2}\right)^2 - \frac{43}{12}}}$$

$$= \frac{1}{\sqrt{3}} \ln\left|x + \frac{3}{2} + \sqrt{\left(x + \frac{3}{2}\right)^2 - \frac{43}{12}}\right| + C$$

by Formula 29 from the table.

Example 23.4.

$$\int \frac{dx}{\sqrt{13 - 9x - 3x^2}}.$$

Completing the square,

$$13 - 9x - 3x^2 = -3\left(x^2 + 3x - \frac{13}{3}\right)$$

$$= -3\left(\left(x + \frac{3}{2}\right)^2 - \frac{79}{12}\right).$$

The integral equals

$$\frac{1}{\sqrt{3}} \int \frac{dx}{\sqrt{\frac{79}{12} - \left(x + \frac{3}{2}\right)^2}} = \frac{1}{\sqrt{3}} \sin^{-1}\frac{x + \frac{3}{2}}{\sqrt{\frac{79}{12}}} + C$$

by Formula 27 from the table.

Exercises 23

1. $\displaystyle\int \cos^2\left(3\theta - \frac{\pi}{12}\right) d\theta$

2. $\displaystyle\int \sin^2\left(\frac{1}{2}\theta + \frac{\pi}{3}\right) d\theta$

3. $\displaystyle\int \cos^6\theta \, d\theta$

4. $\displaystyle\int \sin^6 2\theta \, d\theta$

5. $\displaystyle\int \sin^8\theta \, d\theta$

6. $\displaystyle\int_0^{\frac{\pi}{2}} \sin^{10}\theta \, d\theta$

7. $\displaystyle\int \sin^2\theta \cos^2\theta \, d\theta$

8. $\displaystyle\int \sin^3 2\theta \cos^2 2\theta \, d\theta$

9. $\displaystyle\int x \sin 3x \, dx$

10. $\displaystyle\int x^2 \cos 5x \, dx$

11. $\displaystyle\int \tan^3 x\, dx$

12. $\displaystyle\int \frac{dx}{2 + x^2}$

13. $\displaystyle\int \frac{dx}{42 + 2x^2}$

14. $\displaystyle\int \frac{dx}{x^2 + 2x + 2}$

15. $\displaystyle\int \frac{dx}{(2 + x^2)^2}$

16. $\displaystyle\int \frac{dx}{(1 + 2x^2)^2}$

17. $\displaystyle\int \frac{dx}{(x^2 + 2x + 2)^2}$

18. $\displaystyle\int \frac{dx}{3 - x^2 - 2x}$

19. $\displaystyle\int \frac{dx}{\sqrt{x^2 + 7}}$

20. $\displaystyle\int x\, e^{5x}\, dx$

21. $\displaystyle\int x^5\, e^x\, dx$

22. $\displaystyle\int \frac{dx}{(5 + 2x + 2x^2)^2}$

23. $\displaystyle\int \frac{dx}{(5 + 2x - 2x^2)^2}$

24. $\displaystyle\int \sqrt{6 - 4x - 3x^2}\, dx$

24

Partial Fractions and Integration by Parts

24.1. Partial Fractions

There is a fairly easy way to integrate a fraction of polynomials such as

$$\int \frac{x^4 + 1}{x^3 - x^2 - 14x + 24} \, dx$$

if the denominator factors into distinct linear factors. The idea is that this monster fraction has come from adding together simpler fractions. We want to decompose this large fraction back into those simpler "partial fractions," which should be easier to integrate. The steps in integrating any such fraction

$$\int \frac{P(x)}{Q(x)} \, dx$$

are as follows:

1. If the degree of the numerator $P(x)$ is greater than or equal to the degree of the denominator $Q(x)$, simplify by long division.

2. Factor the denominator $Q(x)$ into distinct linear factors: $Q(x) = (x - a_1)(x - a_2) \ldots (x - a_n)$.

3. Decompose

$$\frac{P(x)}{Q(x)} = \frac{P(x)}{(x - a_1) \ldots (x - a_n)}$$

$$= \frac{b_1}{x - a_1} + \frac{b_2}{x - a_2} + \ldots + \frac{b_n}{x - a_n}.$$

Actually b_1 can be obtained by considering

$$\frac{P(x)}{Q(x)} = \frac{P(x)}{(x - a_1)(x - a_2) \ldots (x - a_n)},$$

covering up $x - a_1$ with your finger, and then plugging in $x = a_1$ to get

$$b_1 = \frac{P(a_1)}{(a_1 - a_2) \ldots (a_1 - a_n)}.$$

Similarly find $b_2, \ldots b_n$.

4. Now the integral is just

$$\int \frac{b_1}{x - a_1} \, dx + \int \frac{b_2}{x - a_2} \, dx + \ldots + \int \frac{b_n}{x - a_n} \, dx$$

$$= b_1 \ln |x - a_1| + \ldots + b_n \ln |x - a_n| + C.$$

Example 24.1.

$$\int \frac{x^4 + 1}{x^3 - x^2 - 14x + 24} \, dx.$$

Step 1. Long division.

$$
\begin{array}{r}
x +1 \\
x^3 - x^2 - 14x + 24 \overline{\smash{\big)}\, x^4 \quad +0x^3 \quad +0x^2 \quad +0x \quad +1} \\
\underline{x^4 \quad -x^3 \quad -14x^2 \quad +24x} \\
x^3 \quad +14x^2 \quad -24x \quad +1 \\
\underline{x^3 \quad -x^2 \quad -14x \quad +24} \\
15x^2 \quad -10x \quad -23
\end{array}
$$

Therefore our integrand is

$$
x + 1 + \frac{15x^2 - 10x - 23}{x^3 - x^2 - 14x + 24}.
$$

Step 2. Factor the denominator $Q(x) = x^3 - x^2 - 14x + 24$. Let's try to guess a root. Since the good guesses must divide 24, try $\pm 1, \pm 2, \pm 3, \pm 6, \pm 12, \pm 24$. $Q(1) = 10$, so $x - 1$ is not a factor. $Q(-1) = 36$ so $x + 1$ is not a factor. $Q(2) = 0$, so $x - 2$ is a factor.

$$
\begin{array}{r}
x^2 \quad +x \quad -12 \\
x - 2 \overline{\smash{\big)}\, x^3 \quad -x^2 \quad -14x \quad +24} \\
\underline{x^3 \quad -2x^2} \\
x^2 \quad -14x \\
\underline{x^2 \quad -2x} \\
-12x \quad +24 \\
\underline{-12x \quad +24}
\end{array}
$$

The remaining $x^2 + x - 12$ factors as $(x - 3)(x + 4)$. Thus $Q(x) = (x - 2)(x - 3)(x + 4)$, and our integrand is

$$
x + 1 + \frac{15x^2 - 10x - 23}{(x - 2)(x - 3)(x + 4)}.
$$

Step 3. Decomposition.

$$\frac{15x^2 - 10x - 23}{(x-2)(x-3)(x+4)} = \frac{b_1}{x-2} + \frac{b_2}{x-3} + \frac{b_3}{x+4}$$

$$b_1 = \frac{15 \cdot 2^2 - 10 \cdot 2 - 23}{(2-3)(2+4)} = \frac{17}{-6}.$$

$$b_2 = \frac{15 \cdot 3^2 - 10 \cdot 3 - 23}{(3-2)(3+4)} = \frac{82}{7}.$$

$$b_3 = \frac{15 \cdot (-4)^2 - 10 \cdot (-4) - 23}{(-4-2)(-4-3)} = \frac{257}{42}.$$

Step 4. The integral of $x + 1$ is $\frac{1}{2}x^2 + x$, so the whole integral is

$$\frac{1}{2}x^2 + x - \frac{17}{6}\ln|x-2| + \frac{82}{7}\ln|x-3| + \frac{257}{42}\ln|x+4| + \text{C}.$$

Remark. The method of partial fractions can be generalized to deal with nonlinear factors of Q and repeated roots.

24.2. Integration by Parts

Unfortunately, the integral of a product of two functions is not just the product of the integrals, just as the derivative of a product is not just the product of the derivatives, but a sum of two parts given by the Product Rule. Reversing the Product Rule for differentiation yields a rule of integration of products "by parts." Since by the Product Rule for differentiation, $d(uv) = u\,dv + v\,du$,

$$\int u\,dv = uv - \int v\,du.$$

Example 24.2. $\int \ln x\,dx$. Let $u = \ln x$, $dv = dx$. Then

$$du = \frac{dx}{x} \quad \text{and} \quad v = x.$$

Hence
$$\int \ln x \, dx = x \ln x - \int dx = x \ln x - x + C.$$

Advice on integration by parts. Use integration by parts to integrate products of functions, especially x^n, e^{ax}, $\ln x$, trigonometric functions, and inverse trigonometric functions. Take u to be the factor you would like to replace by its derivative: possibly e^{ax} or a trig function, better x^n (since it gets simpler by differentiating), and especially $\ln x$ and inverse trig functions (which get much simpler by differentiating). Take dv to be the factor you can integrate: often $e^{ax} \, dx$, $\sin ax \, dx$, $\cos ax \, dx$, or simply dx; sometimes $x^n \, dx$.

Theoretical value. Integration by parts has great theoretical value in mathematics. For example, the very definition of generalized functions or *distributions* in more advanced mathematics depends on integration by parts. Exercises 21 and 22 will give you a glimpse of implications of integration by parts for all functions $f(x)$.

Reduction formulas. Actually the reduction formulas in the tables are obtained by integration by parts. For example consider
$$\int \cos^m x \, dx.$$

Let $u = \cos^{m-1} x$ and $dv = \cos x \, dx$. Then
$$du = -(m-1) \cos^{m-2} x \sin x \, dx, \quad \text{and} \quad v = \sin x.$$

Thus
$$\int \cos^m x \, dx = \cos^{m-1} x \sin x + (m-1) \int \cos^{m-2} x \sin^2 x \, dx$$
$$= \cos^{m-1} x \sin x + (m-1) \int \cos^{m-2} x (1 - \cos^2 x) \, dx,$$

$$(1 + (m-1)) \int \cos^m x \, dx = \cos^{m-1} x \sin x + (m-1) \int \cos^{m-2} x \, dx,$$
$$\int \cos^m x \, dx = \frac{1}{m} \cos^{m-1} x \sin x + \frac{m-1}{m} \int \cos^{m-2} x \, dx.$$

This is the special case of Table Formula 17(2) when $n = 0$.

Exercises 24

1. $\displaystyle\int \frac{dx}{x^2 - 1}$

2. $\displaystyle\int \frac{5dx}{x^2 - 3}$

3. $\displaystyle\int \frac{xdx}{x^2 - 3x - 4}$

4. $\displaystyle\int \frac{x^3 dx}{x^2 - 3x - 4}$

5. $\displaystyle\int \frac{(x^2 - 3)\,dx}{x^3 - 6x^2 + 5x + 12}$

6. $\displaystyle\int \frac{x^3 - x^2 + x + 1}{x^3 + 2x^2 - 13x + 10}\,dx$

7. $\displaystyle\int \frac{dx}{x^2 + 5x + 1}$

8. $\displaystyle\int \frac{dx}{(x - 1)(x - 2)(x - 3)(x - 4)}$

9. $\displaystyle\int xe^x\,dx$

10. $\displaystyle\int x\,e^{5x}\,dx$

11. $\displaystyle\int x\,\cos 2x\,dx$

12. $\displaystyle\int x\,\sin \frac{1}{2}x\,dx$

13. $\displaystyle\int x^2 \sin x \, dx$

14. $\displaystyle\int x^3 \cos 5x \, dx$

15. $\displaystyle\int e^x \cos x \, dx$

16. $\displaystyle\int e^x \sin x \, dx$

*17. $\displaystyle\int x e^x \cos x \, dx$

18. $\displaystyle\int \tan^{-1} x \, dx$

19. $\displaystyle\int \sin^{-1} x \, dx$

20. $\displaystyle\int x \sin x^2 \, dx$

21. Prove that if f is twice differentiable and $f'(1) = 0$, then

$$\int_0^1 x f''(x) \, dx = f(0) - f(1).$$

Check for $f(x) = \cos \pi x$.

22. Prove that if f is twice differentiable and $f(0) = f(\pi) = 0$, then

$$\int_0^\pi f''(x) \sin x \, dx = -\int_0^\pi f(x) \sin x \, dx.$$

25

Numerical Methods

Unfortunately, many integrands such as

$$\int \sin x^2 \, dx$$

have no antiderivative in terms of functions known to humanity. To approximate the integral numerically, one could go back to the definition and use Riemann sums. Certain other methods give more accuracy faster. For example, *Simpson's Rule* comes from approximating the integrand by pieces of parabolas.

25.1 Simpson's Rule

To estimate

$$\int_a^b f(x) \, dx,$$

divide the interval $[a, b]$ *into an even number n pieces of length*

$$\Delta x = \frac{b - a}{n},$$

Table 25.1. The data for estimating the area of the unit circle by Simpson's Rule with $n = 4$, $\Delta x = 1/4$.

i	x_i	y_i
0	0	4
1	$\frac{1}{4}$	$\sqrt{15}$
2	$\frac{1}{2}$	$\sqrt{12}$
3	$\frac{3}{4}$	$\sqrt{7}$
4	1	0

marked off by $x_0 = a_1$, $x_1 = a + \Delta x$, $x_2 = a + 2\Delta x$, $\ldots$, $x_n = b$. Let $y_0 = f(x_0), \ldots y_n = f(x_n)$. Then

$$\int_a^b f(x)\,dx \approx \frac{\Delta x}{3}(y_0 + 4y_1 + 2y_2 + 4y_3 + 2y_4 + \ldots + 4y_{n-1} + y_n).$$

If $|f''''| \le$ C for $a \le x \le b$, then

$$|\text{error}| \le C\,\frac{b-a}{180}(\Delta x)^4. \tag{25.1}$$

Example 25.1. As an illustration, let's approximate the area of the unit circle.

$$\text{Area} = \int_{x=-1}^1 \left(\sqrt{1-x^2} - \left(-\sqrt{1-x^2}\right)\right)dx = \int_0^1 4\sqrt{1-x^2}\,dx,$$

which we know is exactly π. For $n = 4$, $\Delta x = \frac{1}{4}$. The data is assembled in Table 25.1.

$$\int_0^4 \sqrt{1-x^2}\,dx \approx \frac{1}{3 \cdot 4}\left(4 + 4\sqrt{15} + 2\sqrt{12} + 4\sqrt{7} + 0\right) = 3.08\ldots.$$

To estimate the error, compute

$$
\begin{aligned}
f'(x) &= -4x(1-x^2)^{-\frac{1}{2}}, \\
f''(x) &= -4(1-x^2)^{-\frac{1}{2}} - 4x^2(1-x^2)^{-\frac{3}{2}} = -4(1-x^2)^{-\frac{3}{2}},
\end{aligned}
$$

$$f'''(x) = -12x(1-x^2)^{-\frac{5}{2}},$$

$$f''''(x) = -12(1-x^2)^{-\frac{5}{2}} - 60x^2(1-x^2)^{-\frac{7}{2}}.$$

We want a constant C such that for $0 \le x \le 1$, $|f''''(x)| \le C$. We are in trouble: as $x \to 1$, $f''''(x)$ blows up! We get no error estimate. Had we had $0 \le x \le \frac{1}{2}$, then $1/(1-x^2)$ hits its biggest value when the denominator is smallest, i.e. when $x = \frac{1}{2}$. Thus

$$\left| \frac{1}{1-x^2} \right| \le \left| \frac{1}{1-(\frac{1}{2})^2} \right| = \frac{1}{\frac{3}{4}} = \frac{4}{3},$$

and

$$|f''''(x)| \le 12\left(\frac{4}{3}\right)^{\frac{5}{2}} + 60 \cdot \frac{1}{4} \cdot \left(\frac{4}{3}\right)^{\frac{7}{2}} \le 66.$$

In practice, one takes n larger and larger until the last decimal place with which you are concerned settles down. For this problem, $n = 10$ yields about 3.13, and $n = 100$ yields about 3.14113 (while $\pi = 3.14159\ldots$), so that for this problem the convergence is unusually slow.

Computers. Most computers have Simpson's Rule or superior methods as part of their software, so that you can just give them the integral and they give you as accurate a numerical answer as you like.

Exercises 25

Use Simpson's Rule with $n = 4$ to estimate the following integrals. When possible, compute the exact answer by antidifferentiating.

1. $\displaystyle\int_1^2 \frac{dx}{x}$

2. $\displaystyle\int_0^2 x\,dx$

3. $\int_0^2 x^2\,dx$

4. $\int_0^2 x^3\,dx$

5. $\int_0^2 x^4\,dx$

6. $\int_0^\pi \sin^2 x\,dx$

7. $\int_0^2 e^{-x^2}\,dx$

Compute to within .001 by trying different values of n.

8. $\ln 2 = \int_1^2 \dfrac{dx}{x}$

9. $\ln 10 = \int_1^{10} \dfrac{dx}{x}$

10. Find a value of n you can *prove* works for Problem 8, using Equation 25.1.

11. Find a value of n you can *prove* works for Problem 9, using Equation 25.1.

12. Use Equation 25.1 to prove that Simpson's Rule has to give the exact answer in Problems 2–4 (but not 5).

26

Review Problems

1. $\displaystyle\int_{\frac{\pi}{6}}^{\frac{\pi}{3}} \cos x \, dx$

2. $\displaystyle\int \cos^2 5x \, \sin 5x \, dx$

3. $\displaystyle\int x e^x \, dx$

4. $\displaystyle\int x e^{x^2} \, dx$

5. $\displaystyle\int_9^{9e} \frac{dx}{x}$ (Simplify answer.)

6. $\displaystyle\int \tan x \, \cos x \, dx$

7. $\displaystyle\int_1^2 \left(\frac{1}{2}\right)^x dx$

8. $\int \dfrac{dx}{x^2 + 4}$

9. $\int \dfrac{x\,dx}{x^2 + 4}$

10. $\int \dfrac{dx}{\sqrt{4 - x^2}}$

11. $\int \dfrac{x\,dx}{\sqrt{4 - x^2}}$

12. $\int \dfrac{e^{\tan^{-1} x}\,dx}{1 + x^2}$

13. a. $\int x(x + 1)(x + 2)\,dx$

    b. $\int_0^2 xe^{x^2}\,dx$

    c. $\int x \sin x\,dx$

    d. $\int_0^{\sqrt{5}} \dfrac{dx}{x^2 + 5}$

    e. $\int \dfrac{x\,dx}{x^2 + 5}$

    f. $\int \sec^2 x \tan x\,dx$

    g. $\int \arcsin x\,dx$

    h. $\int_{\frac{\pi}{4}}^{\frac{\pi}{2}} (x \cos x)\,dx$

14. $\int \dfrac{e^x}{2 + e^{2x}}\,dx$ or $\int \dfrac{e^x}{\sqrt{1 - e^{2x}}}\,dx$

15. $\displaystyle\int \frac{\ln x}{x}\,dx$

16. $\displaystyle\int x(\ln x)^2\,dx$ (parts twice)

17. $\displaystyle\int \tan^2 x \sec^2 x\,dx$

18. $\displaystyle\int 5x + \frac{2}{\sqrt{x}}\,dx$

19. $\displaystyle\int \frac{3x}{25 + x^2}\,dx$

20. $\displaystyle\int \frac{1}{2x + 3}\,dx$

21. $\displaystyle\int \arctan x\,dx$

22. $\displaystyle\int \frac{\arctan x}{1 + x^2}\,dx$

23. $\displaystyle\int \frac{4}{\sqrt{9 - x^2}}\,dx$

24. $\displaystyle\int x e^{5x}\,dx$

25. $\displaystyle\int \frac{4 + x^2}{x^3}\,dx$

26. $\displaystyle\int \frac{x + 1}{x}\,dx$

27. $\displaystyle\int x e^{x^2}\,dx$

28. $\int x \ln(x+3)\,dx$

29. Compute: $e^{3\ln\frac{1}{2}}$, $\ln\left(e^{\frac{1}{5}}\right)^3$, $e^{\ln\left(\frac{1}{2}\right)^3}$.

30. Solve each of the following for x:

    a. $\ln(2e^x) + x = 1$.

    b. $\cos(\arctan(2x)) = \frac{1}{5}$.

31. Find the area of the region bounded by:

    a. the parabolas $y = 6x - x^2$ and $y = x^2 - 2x$.

    b. the parabolas $y = x^2 - 8$ and $y = -2x^2 + 3x + 10$.

32. Do the integral or at least give the method if any.

    a. $\int \dfrac{x\,dx}{\sqrt{3 - x^2}}$

    b. $\int \dfrac{dx}{\sqrt{3 - x^2}}$

    c. $\int \dfrac{dx}{\sqrt{1 - 3x^2}}$

    d. $\int \dfrac{dx}{\sqrt{3 - x}}$

    e. $\int \dfrac{dx}{\sqrt{3 - x^3}}$

    f. $\int \dfrac{x^2\,dx}{\sqrt{3 - x^2}}$

    g. $\int \dfrac{x^2\,dx}{\sqrt[4]{3 - x^3}}$

33. Do the integral or at least give the method if any.

a. $\int \sin 3x^2 \, dx$

b. $\int x \sin 3x^2 \, dx$

c. $\int \sin^2 3x \, dx$

d. $\int \sin^2 3x \cos 3x \, dx$

e. $\int \sin^2 3x \cos x \, dx$

f. $\int \sin^2 3x^2 \cos 3x^2 \, dx$

g. $\int x \sin^2 3x^2 \cos 3x^2 \, dx$

h. $\int x \sin^2 x \, dx$

34. $\int_{x=\frac{2}{3}}^{1} \frac{\sin \pi x}{1 + 4 \cos \pi x} \, dx$

35. Simplify.

a. $\ln e^x$

b. $\cos \cos^{-1} .83$

c. $\sin \sin^{-1} 1.7$

d. $\cos^{-1} \cos \frac{5\pi}{8}$

e. $\sin^{-1} \sin \frac{5\pi}{8}$

f. $\dfrac{d}{dx} \csc 2\pi x$

g. $\dfrac{d}{dx} \cos^{-1} x$

36. Identify possible methods.

a. $\displaystyle\int \frac{dx}{x^2 - x - 42}$

b. $\displaystyle\int \frac{dx}{(x - \frac{1}{2})^2 - 42\frac{1}{4}}$

c. $\displaystyle\int \frac{dx}{x^2 - x + 42}$

d. $\displaystyle\int \frac{dx}{(x - \frac{1}{2})^2 + 41\frac{3}{4}}$

e. $\displaystyle\int \frac{(2x - 1)\, dx}{x^2 - x - 42}$

f. $\displaystyle\int \frac{2x\, dx}{x^2 - 4x - 42}$

g. $\displaystyle\int \frac{dx}{x^2 + 6x + 9}$

h. $\displaystyle\int \frac{dx}{\sqrt{6 - x^2}}$

i. $\displaystyle\int \frac{dx}{\sqrt{6 + x^2}}$

j. $\displaystyle\int \frac{dx}{\sqrt{x^2 - 6}}$

k. $\displaystyle\int \frac{x^5}{x^2 - 1}\, dx$

l. $\displaystyle\int \frac{x^5}{x^2 + 1}\, dx$

m. $\displaystyle\int_0^1 e^{-x^2}\, dx$

n. $\displaystyle\int 4x \sin 4x\, dx$

37. Give the derivative and the indefinite integral.

    a. $\tan x$

    b. $\sec x$

    c. $\sec^2 x$

    d. $\tan^2 x$

    e. $(a^2 - x^2)^{-\frac{1}{2}}$

    f. $\sin^{-1} x$

    g. 2^x

    h. $\ln ax$

38. Give the solution or at least the method if any.

    a. $\displaystyle\int \frac{dx}{x^2 - 5}$

    b. $\displaystyle\int \frac{dx}{x^2 + 5}$

    c. $\displaystyle\int \frac{x\,dx}{x^2 - 5}$

    d. $\displaystyle\int \frac{x\,dx}{x^2 + 5}$

    e. $\displaystyle\int \frac{x^2\,dx}{x^2 - 5}$

    f. $\displaystyle\int \frac{x^2\,dx}{x^2 + 5}$

    g. $\displaystyle\int \frac{dx}{4x^2 + x + 3}$

    h. $\displaystyle\int x \sec^2 x\,dx$

39. Identify the pattern: πr^2, $\frac{4}{3}\pi r^3$, $\frac{1}{2}\pi^2 r^4$, $\frac{8}{15}\pi^2 r^5$, $\frac{1}{6}\pi^3 r^6$, $\ldots$.

40. Simplify.

    a. $e^{\ln e^x}$

    b. $\left(\dfrac{4}{9}\right)^{-\frac{3}{2}}$

    c. $(i)^{\frac{1}{2}}$

    d. $(i)^i$

41. When do you use these formulas?

    a. $y = y_0 e^{kt}$

    b. $y = y_0 \left(\dfrac{1}{2}\right)^{\frac{t}{h}}$

    c. $ds = \sqrt{\dot{x}^2 + \dot{y}^2}\, dt$

    d. $dV = (\pi r_2{}^2 - \pi r_1{}^2)\, dx$

    e. $\displaystyle\int u\, dv = uv - \int v\, du$

42. Discuss.

    a. Riemann integrals and sums

    b. Fundamental Theorem of Calculus

    c. Simpson's Rule

Part III

Infinite Series

27

Infinite Series (Sums)

The ancient Greeks were confused about the initially astonishing fact that *an infinite series can have a finite sum.* For example,

$$\frac{1}{2} + \frac{1}{4} + \frac{1}{8} + \frac{1}{16} + \ldots = 1. \tag{27.1}$$

The first two terms add up to 3/4, the first three add up to 7/8, the first four add up to 15/16. As the number of terms increases the subtotals or *partial sums* converge to 1. We say that the whole series *converges* or sums to 1 and write the equality given in Equation 27.1. The subtotals do not need to ever reach 1.

Since this series can be written

$$\frac{1}{2^1} + \frac{1}{2^2} + \frac{1}{2^3} + \frac{1}{2^4} + \cdots,$$

we abbreviate it by

$$\sum_{n=1}^{\infty} \frac{1}{2^n},$$

"the sum of all $\frac{1}{2^n}$ as n goes from 1 to ∞," using the capital Greek letter Σ (Sigma).

Any series $a_1 + a_2 + a_3 + \ldots$ can be abbreviated

$$\sum_{n=1}^{\infty} a_n.$$

An infinite series

$$\sum_{n=1}^{\infty} a_n$$

can do one of four things:

1. *converge* to a finite number, such as Equation 27.1,

2. *diverge* to $+\infty$, such as

$$1 + 2 + 3 + 4 + \ldots = \sum_{n=1}^{\infty} n \to +\infty,$$

3. *diverge* to $-\infty$, such as

$$1 - 3 + 2 - 4 + 3 - 5 + 4 - 6 + \ldots \to -\infty$$

(where the partial sums are:

$1, -2, 0, -4, -1, -6, -2, -8, \ldots \to -\infty)$,

4. *diverge by oscillation*, such as

$1 - 1 + 1 - 1 + 1 - 1 + 1 \ldots$ or $1 - 1 + 2 - 2 + 3 - 3 + \ldots.$

27.1. Zeno's Paradox

Zeno perplexed his fellow Greeks with a paradoxical argument that Achilles, once behind, could never catch up with a tortoise in a race. He argued that each time Achilles advanced to the tortoise's position, the tortoise would have moved a little bit farther. Zeno was right that from this perspective it would take infinitely many periods of time for Achilles to catch up. His mistake was in not understanding that such an infinite series of times can converge, can add up to a finite total amount of time in which Achilles overtakes the tortoise.

27.2. Convergence of Series

In general we like series to converge. Of course, for a series to converge, the terms must go to 0: $a_n \to 0$. Unfortunately, even if the terms go to 0, the series may not converge. For example, the *harmonic series*

$$\sum_{n=1}^{\infty} \frac{1}{n} = 1 + \frac{1}{2} + \frac{1}{3} + \frac{1}{4} + \frac{1}{5} + \frac{1}{6} + \frac{1}{7} + \frac{1}{8} + \cdots$$

$$\geq 1 + \frac{1}{2} + \left(\frac{1}{4} + \frac{1}{4}\right) + \left(\frac{1}{8} + \frac{1}{8} + \frac{1}{8} + \frac{1}{8}\right) + \cdots$$

$$= 1 + \frac{1}{2} + \frac{1}{2} + \frac{1}{2} + \cdots$$

$$\to +\infty$$

even though $\frac{1}{n} \to 0$. Therefore you can only conclude that a series does *not* converge if its terms do *not* go to 0.

By the way,

$$\sum_{n=1}^{\infty} \frac{1}{n^2} = 1 + \frac{1}{2^2} + \frac{1}{3^2} + \cdots$$

does converge (to $\frac{\pi^2}{6}$). In fact,

27.3. The p-Test

$$\sum_{n=1}^{\infty} \frac{1}{n^p} \quad \textit{converges if and only if } p > 1.$$

The sum is called $\zeta(p)$, where the Greek letter ζ (zeta) stands for the famous Riemann zeta function. If this function were well understood, most of the open questions in number theory would be solved. No one even knows exactly the value of

$$\zeta(3) = 1 + \frac{1}{2^3} + \frac{1}{3^3} + \cdots.$$

27.4. Geometric Series

Along with the series

$$\sum_{n=1}^{\infty} \frac{1}{n^p} \ ,$$

the most important series are geometric series. Each term in a geometric series is obtained by multiplying the previous one by a fixed ratio r.

$$\sum_{n=1}^{\infty} \frac{1}{2^n} = \frac{1}{2} + \frac{1}{4} + \frac{1}{8} + \dots$$

is a geometric series with a ratio $r = \frac{1}{2}$.

$$\sum_{n=1}^{\infty} (-1)^{n+1} = 1 - 1 + 1 - 1 \dots$$

is a geometric series with a ratio $r = -1$. If the starting term of a geometric series is a_0, then the next is $a_0 \cdot r$, the next $a_0 \cdot r^2$, and the whole series is

$$\sum_{n=0}^{\infty} a_0 \cdot r^n.$$

27.5. Geometric Series Test

A geometric series

$$\sum_{n=0}^{\infty} a_0 \, r^n \quad converges \ to \quad \frac{a_0}{1 - r}$$

if $|r| < 1$ and diverges if $|r| \geq 1$.

For example,

$$\sum_{n=1}^{\infty} \frac{1}{2^n} \quad \text{converges to} \quad \frac{\frac{1}{2}}{1 - \frac{1}{2}} = 1,$$

and

$$\frac{1}{5} + \frac{1}{15} + \frac{1}{45} + \frac{1}{135} + \dots \quad \text{converges to} \quad \frac{\frac{1}{5}}{1 - \frac{1}{3}} = 0.3.$$

The formula

$$\frac{a_0}{1-r}$$

for the sum A can be discovered as follows:

$$
\begin{aligned}
A &= a_0 &+ a_0 r + a_0 r^2 + \dots \\
-\ (rA &= & a_0 r + a_0 r^2 + \dots)
\end{aligned}
$$

$$(1-r)A = a_0$$

and hence

$$A = \frac{a_0}{1-r}.$$

A similar calculation for partial sums would be the main step in a proof of convergence for $|r| < 1$. Of course for $|r| \geq 1$, the terms do not go to 0, so the series must diverge.

The following test for convergence is simple, perhaps obvious, but very useful.

27.6. The Comparison Test

Suppose $|b_n| \leq a_n$. *If* $\sum a_n$ *converges, then* $\sum b_n$ *converges. If* $\sum b_n$ *diverges, then* $\sum a_n$ *diverges.*
    For example,

$$\sum_{n=1}^{\infty} \frac{1}{n \cdot 2^n} = \frac{1}{1 \cdot 2} + \frac{1}{2 \cdot 4} + \frac{1}{3 \cdot 8} + \dots$$

converges by comparison with

$$\sum \frac{1}{2^n},$$

although we do not yet know to what it converges. It follows of course that

$$\sum_{n=1}^{\infty} \frac{6}{n \cdot 2^n} \quad \text{converges (to six times as much as} \quad \sum_{n=1}^{\infty} \frac{1}{n \cdot 2^n}).$$

On the other hand,

$$\sum_{n=1}^{\infty} \frac{1}{\sqrt{n}} = 1 + \frac{1}{\sqrt{2}} + \frac{1}{\sqrt{3}} + \dots$$

diverges by comparison with $\sum \frac{1}{n}$ (or by the p-test (27.3) since $p = \frac{1}{2} < 1$). It follows of course that for any positive number a

$$\sum_{n=1}^{\infty} \frac{a}{\sqrt{n}}$$

diverges.

The following is the most important test for convergence, especially good for powers and factorials. It uses the Greek letter ρ (rho).

27.7. The Ratio Test

$$Let \quad \rho = \lim_{n\to\infty} \left| \frac{a_{n+1}}{a_n} \right|.$$

If $\rho < 1$, $\sum a_n$ *converges.*
If $\rho > 1$, $\sum a_n$ *diverges.*
If $\rho = 1$ *or does not exist, you cannot tell from this test.*

For example, consider

$$\sum_{n=1}^{\infty} \frac{2^n}{n!}.$$

Then

$$\rho = \lim_{n\to\infty} \frac{\frac{2^{n+1}}{(n+1)!}}{\frac{2^n}{n!}}$$

$$= \lim_{n\to\infty} \frac{2^{n+1}}{2^n} \cdot \frac{n!}{(n+1)!}$$

$$= \lim_{n\to\infty} \frac{2}{(n+1)} = 0.$$

Therefore $\sum_{n=1}^{\infty} \frac{2^n}{n!}$ converges. (Actually, it converges to $e^2 - 1$.)

Remark on proof. The proof goes by comparing the series with a geometric series with $r \approx \rho$. In the borderline case $\rho = 1$, the approximation is not fine enough to distinguish between convergence and divergence. For example,

$$\sum \frac{1}{n} \quad \text{and} \quad \sum \frac{1}{n^2}$$

both have $\rho = 1$, but the first diverges while the second converges.

27.8. Remark

Changing the first million terms has no effect on whether or not a series converges, although it may well affect the sum to which it converges.

Exercises 27

Tell whether or not the series converges. Give a reason. If possible, give the sum.

1. $\displaystyle\sum_{n=1}^{\infty} \frac{1}{5^n}$

2. $\displaystyle\sum_{n=1}^{\infty} \frac{100}{\pi^n}$

3. $\displaystyle\sum_{n=1}^{\infty} \frac{1}{\sqrt[3]{n}}$

4. $\displaystyle\sum_{n=1}^{\infty} \frac{1}{n^5}$

5. $\displaystyle\sum_{n=1}^{\infty} \frac{1}{n\sqrt{n}}$

6. $\displaystyle\sum_{n=1}^{\infty} \ln n$

7. $\displaystyle\sum_{n=1}^{\infty} (-1)^n n^2$

8. $\displaystyle\sum_{n=1}^{\infty} \frac{n^2}{2^n}$

9. $\displaystyle\sum_{n=1}^{\infty} \frac{n!}{n^3 e^n}$

10. $\displaystyle\sum_{n=1}^{\infty} \frac{(e^n)^2}{n!}$

11. $\displaystyle\sum_{n=1}^{\infty} \frac{100^n}{n!}$

12. $\displaystyle\sum_{n=1}^{\infty} \frac{n-1}{n^3}$

13. $\displaystyle\sum_{n=1}^{\infty} \frac{1}{(n+1)^2}$

14. $\displaystyle\sum_{n=1}^{\infty} \frac{1}{\left(n-\frac{1}{2}\right)^2}$

15. $\displaystyle\sum_{n=1}^{\infty} \frac{1-e^{-n}}{n^2}$

16. $\displaystyle\sum_{n=1}^{\infty} \frac{1+e^{-n}}{n^2}$

28

Power Series and Taylor Series

Sometimes even transcendental functions can be given exactly by a *power series*—infinite series in powers of x. The five most important examples to memorize are:

$$\frac{1}{1-x} \;=\; 1 + x + x^2 + x^3 + \dots \qquad |x| < 1 \qquad (28.1)$$

$$e^x \;=\; 1 + x + \frac{x^2}{2!} + \frac{x^3}{3!} + \dots \qquad \text{all } x \qquad (28.2)$$

$$\sin x \;=\; x - \frac{x^3}{3!} + \frac{x^5}{5!} + \dots \qquad \text{all } x \qquad (28.3)$$

$$\cos x \;=\; 1 - \frac{x^2}{2!} + \frac{x^4}{4!} + \dots \qquad \text{all } x \qquad (28.4)$$

$$(1+x)^p \;=\; 1 + px + \frac{p(p-1)}{2}\, x^2$$

$$+\; \frac{p(p-1)(p-2)}{3!}\, x^3 + \dots \quad |x| < 1 \qquad (28.5)$$

The first is the already familiar geometric series, with $r = x$, which we know converges to

$$\frac{1}{1-x} \quad \text{when} \quad |x| < 1.$$

The second tells us for example that

$$e^2 = \sum_{n=0}^{\infty} \frac{2^n}{n!} \quad \text{(we agree } 0! = 1\text{)}$$

and explains how I got $e^2 - 1$ as the sum of the series in Section 27.7.

The last is called the binomial series, and has only finitely many terms if p is an integer. For example,

$$(1+x)^3 = 1 + 3x + 3x^2 + x^3$$
$$(1+x)^{\frac{1}{2}} = 1 + \frac{1}{2}x - \frac{1}{8}x^2 + \frac{1}{16}x^3 - \cdots$$

Functions $f(x)$ given by a power series

$$f(x) = \sum_{n=0}^{\infty} a_n x^n \qquad (28.6)$$

are called *real-analytic*. Such a series can be differentiated to any order. Notice that differentiating the $\sin x$ series once yields the $\cos x$ series, and differentiating again yields the series for $-\sin x$.

The coefficients a_n are given by the formula

$$a_n = \frac{f^{(n)}(0)}{n!}, \qquad (28.7)$$

where $f^{(n)}$ denotes the nth derivative of f. This rule may be discovered by differentiating Formula 28.6 n times and then plugging in $x = 0$, so that only the constant term remains. For example, if $f(x) = e^x$, $f'(x) = e^x, \ldots, f^{(n)}(x) = e^x$,

$$a_n = \frac{f^{(n)}(0)}{n!} = \frac{1}{n!},$$

which yields Formula 28.2.

Although all power series could be derived from Formula 28.7, it is much easier to memorize Formulas 28.1–28.5 and derive others by manipulating them: making substitutions, integrating, differentiating. For example,

$$e^{x^3} = 1 + (x^3) + \frac{(x^3)^2}{2!} + \frac{(x^3)^3}{3!} + \ldots$$

$$= 1 + x^3 + \frac{x^6}{2!} + \frac{x^9}{3!} + \ldots.$$

As a second example,

$$\frac{1}{1 - x + x^2} = \frac{1}{1 - (x - x^2)}$$

$$= 1 + (x - x^2) + (x - x^2)^2 + (x - x^2)^3 + \ldots$$

$$= 1 + x + 0x^2 - x^3 + \ldots \qquad (|x - x^2| < 1).$$

As a third example we compute the power series for $f(x) = \tan^{-1}x$. For this function it is easier to start with the derivative:

$$f'(x) = \frac{1}{1 + x^2} = \frac{1}{1 - (-x^2)} = 1 - x^2 + x^4 - x^6 + \ldots \qquad (|x| < 1).$$

Hence

$$\tan^{-1}x = f(x) = x - \frac{x^3}{3} + \frac{x^5}{5} - \frac{x^7}{7} + \ldots + C,$$

where $C = f(0) = \tan^{-1}(0) = 0$. Although we have established this formula only for $|x| < 1$, it remains true for $x = 1$, and yields Leibniz's Formula for π:

$$\frac{\pi}{4} = \tan^{-1}1 = 1 - \frac{1}{3} + \frac{1}{5} - \frac{1}{7} + \ldots$$

$$\pi = \frac{4}{1} - \frac{4}{3} + \frac{4}{5} - \frac{4}{7} + \ldots.$$

28.1. Taylor Series

For $x \approx a$, a power series is more likely to converge quickly if given in powers of $(x - a)$, called the Taylor series:

$$\sum_{n=0}^{\infty} \frac{f^n(a)}{n!} (x - a)^n .$$

In practice it is usually better to make the substitution $u = x - a$ and get back to powers of u. For example, to find the Taylor series for $y = \frac{1}{x}$ about $a = 1$, let $u = x - 1$,

$$
\begin{aligned}
y &= \frac{1}{1 + u} = 1 - u + u^2 - u^3 + \dots & |u| < 1 \\
&= 1 - (x - 1) + (x - 1)^2 - (x - 1)^3 + \dots & |x - 1| < 1
\end{aligned}
$$

Remarks. Even if $f(x)$ has derivatives of all orders, its series given by Formula 28.7 might not converge to $f(x)$. This failure will not occur for most functions you will encounter. It occurs for

$$f(x) = e^{-1/x^2} .$$

All the derivatives vanish at 0, so that $a_0 = a_1 = a_2 = \dots = 0$, the Taylor series is the 0 function, but f is not the 0 function.

Exercises 28

Find the power series by manipulating Formulas 28.1–28.5.

1. e^{2x} (Check your answer using Formula 28.6.)

2. $\sin x^2$ (Check your answer using Formula 28.6.)

3. $\dfrac{1}{1 - x^2}$

4. $\dfrac{1}{(1 - x)^2}$

5. $\ln(1 + x)$ (Hint: what is the series for its derivative?)

6. $e^x \cos x$

   (Compute the terms up to x^5 by multiplying the two series together.)

7. $\tan x$

   (Compute the terms up to x^4 by long division.)

8. $\sin^{-1} x$

   (Compute the terms up to x^4 by applying Formula 28.5 to its derivative.)

9. Compute $\sqrt{1.5}$ to four decimal places using Formula 28.5.

10. Compute e to four decimal places using Formula 28.2.

Part IV

Differential Equations

29

Differential Equations

Many useful equations in biology, economics, and other physical and social sciences involve derivatives. A random differential equation such as

$$e^y \sqrt{1 + (y')^2} + \ln y'' = \sec \sqrt{x^2 + 1}$$

could be hard or impossible to solve for y as a function of x. Fortunately, for a few types of differential equations that arise frequently, solutions have been found. (Of course, computers can solve many more of these equations numerically or graphically. This chapter may help you appreciate later what the computer can do for you.)

Actually you already know how to solve two sorts of differential equations. First, a differential equation of the simple form

$$\frac{dy}{dx} = f(x)$$

is solved by integrating. For example, if

$$\frac{dy}{dx} = \tan x, \quad \text{then}$$
$$y = -\ln|\cos x| + C.$$

Second, a differential equation of the form

$$\frac{dy}{dx} = ky$$

has as its solutions the exponential functions $y = y_0 e^{kx}$. These two simple kinds of differential equations are both examples of "separable differential equations," our first new topic.

Note that the solution to a differential equation is a function, not just a number.

29.1. Separable Differential Equations

A separable differential equation is one of the form

$$\frac{dy}{dx} = f(x)\,g(y) \quad \text{or} \quad \frac{dy}{dx} = \frac{f(x)}{g(y)} \quad \text{or} \quad \frac{dy}{dx} = \frac{g(y)}{f(x)}.$$

You can solve it by separating the x and y stuff to the two sides and integrating each side.

Example 29.1. Solve

$$y' = xy^2 + x.$$

Solution. First note that the right-hand side can be rewritten as a function of x times a function of y:

$$\frac{dy}{dx} = x(y^2 + 1).$$

Having also rewritten y' as $\dfrac{dy}{dx}$, we can easily separate the x and y stuff to the two sides:

$$\frac{dy}{1 + y^2} = x\,dx.$$

Now integrate:

$$\int \frac{dy}{1 + y^2} = \int x\,dx$$

$$\tan^{-1} y + C_1 \;=\; \frac{1}{2}x^2 + C_2$$

$$\tan^{-1} y \;=\; \frac{1}{2}x^2 + C \qquad\qquad (29.1)$$

$$y \;=\; \tan\left(\frac{1}{2}x^2 + C\right). \qquad\qquad (29.2)$$

Checking, we compute

$$\frac{dy}{dx} \;=\; x\sec^2\left(\frac{1}{2}x^2 + C\right) = x\left(1 + \tan^2\left(\frac{1}{2}x^2 + C\right)\right)$$

$$=\; x(y^2 + 1).$$

Sometimes it is hard to solve explicitly for y as in Equation 29.2 and one is satisfied with an implicit solution as in Equation 29.1.

29.2. First Order Linear Differential Equations

It is very common to encounter a differential equation of the form

$$y' + p(x)y = q(x). \qquad\qquad (29.3)$$

It is called *first order* because only the *first* derivative occurs. It is called *linear* (in y) because y' and y appear only as *linear* terms, which means that they are raised to the first power only, no y^2, no $1/y$, no yy', no $\cos y'$, etc. The functions $p(x)$ and $q(x)$ of x can be anything.

Memorize this formula for the solution. *Let*

$$\rho(x) = e^{\int p(x)dx} \quad (\textit{no need for constant of integration}).$$

Then

$$y = \rho^{-1}\left(\int \rho q \, dx + C\right). \qquad\qquad (29.4)$$

Note the difference between the familiar letter p and the Greek letter ρ called "rho."

Example 29.2. Solve $xy' + y - x\cos x = 0$.

Solution. By dividing by x we can make the differential equation match up exactly with form 29.3 for a first order, linear differential equation:

$$y' + \frac{1}{x}y = \cos x,$$

where $p(x) = \frac{1}{x}$ and $q(x) = \cos x$. Hence

$$\rho(x) = e^{\int p(x)dx} = e^{\int \frac{1}{x}dx} = e^{\ln|x|} = |x| = \pm x.$$

Since it is apparent from Formula 29.4 that multiplying ρ by a constant (such as -1) only changes the constant C, we may take $\rho(x) = x$. Then

$$y = x^{-1}\left(\int x\cos x\, dx + C\right).$$

Integrating by parts with $u = x$, $dv = \cos x\, dx$, $du = dx$, $v = \sin x$, yields

$$
\begin{aligned}
y &= x^{-1}\left(x\sin x - \int \sin x\, dx + C\right)\\
&= x^{-1}\left(x\sin x + \cos x + C\right)\\
&= \sin x + x^{-1}\cos x + Cx^{-1}.
\end{aligned}
$$

One can check this answer by plugging it back into the original differential equation. Notice that for general differential equations the constant of integration often gets mixed in with other functions.

29.3. Initial Value Problems (IVP)

To evaluate the constant of integration C, you need some additional data, such as the initial value of y when $x = 0$ or $x = a$. In the previous example, if you knew that

$$y\left(\frac{\pi}{3}\right) = 1,$$

you could plug that information into the solution and solve for C:

$$1 = \sin\frac{\pi}{3} + \frac{3}{\pi}\cos\frac{\pi}{3} + C\frac{3}{\pi}$$

$$1 - \frac{\sqrt{3}}{2} - \frac{3}{\pi} \cdot \frac{1}{2} = C\frac{3}{\pi}$$

$$C = \frac{\pi}{3} - \frac{\pi\sqrt{3}}{6} - \frac{1}{2}$$

$$= \frac{1}{6}\left(2\pi - \sqrt{3}\pi - 3\right).$$

Thus

$$y = \sin x + x^{-1}\cos x + \frac{1}{6}\left(2\pi - \sqrt{3}\pi - 3\right)x^{-1}.$$

29.4. Interpreting Differential Equations

A differential equation $y' = F(x, y)$ gives the rate of change of y, i.e., the slope of the graph of $y(x)$ at any point (x, y). Here $F(x, y)$ is a formula in terms of x and y for the derivative y', not to be confused with a formula $f(x)$ for y as a function of x. When F is positive, y is increasing; when F is negative, y is decreasing.

Figure 29.1 indicates the slopes prescribed by the differential equation $y' = F(x, y) = y^2 - x$. Figure 29.2 sketches some graphs with the prescribed slopes, i.e., graphs of solutions of the differential equation.

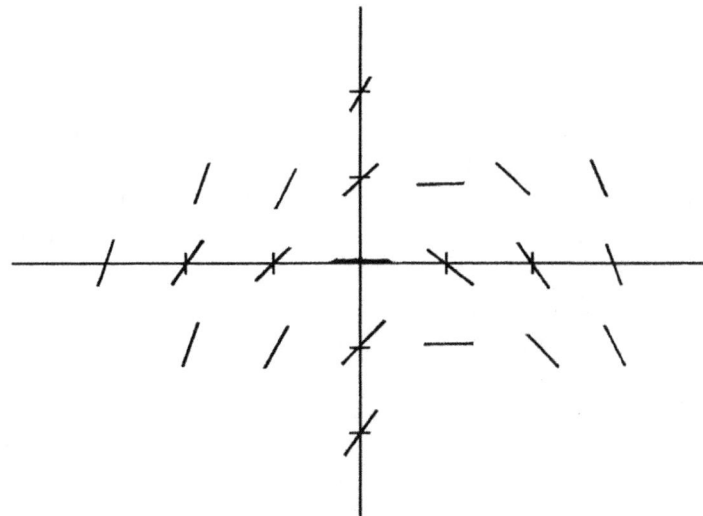

Figure 29.1. The slopes prescribed by the differential equation $y' = F(x, y) = y^2 - x$.

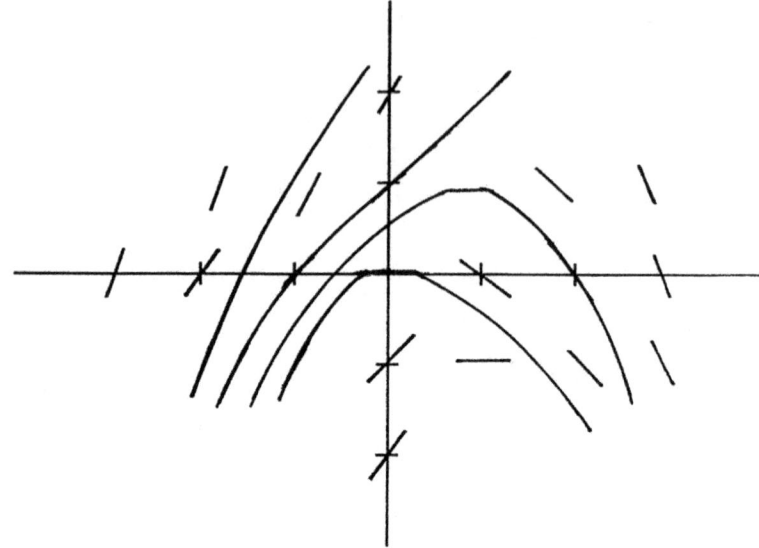

Figure 29.2. Some graphs of solutions of the differential equation $y' = F(x,y) = y^2 - x$.

Note how knowing an initial value $y(0)$ determines which solution is found. I do not know how to solve this differential equation exactly.

29.5. Mixing Problems

Mixing problems involve changing concentrations of solvents or pollutants in reservoirs of water. The concentration c of pollutant (in pounds per gallon) equals the amount y of pollutant (in pounds) divided by the volume V of the reservoir (in gallons). The trick to solving mixing problems is to figure out first the volume $V(t)$ (often constant), then the amount $y(t)$, and thus obtain the concentration $c(t) = y(t)/V(t)$. The following example considers how fast a water body cleans up if the entering pollution decreases rapidly, perhaps due to a new environmental protection law.

Example 29.3.   Part A.   Suppose water flows in and out of Boston's Charles River Basin (of total volume V) at a constant rate of r gallons/year. Suppose that the entering water has pollutant concentration $c_1 e^{-\alpha t}$ pounds/gallon. (This Greek letter α ("alpha") is just a constant.) Find the equation for pollutant concentration $c(t)$

of the basin (assuming uniform pollutant concentration throughout the basin).

Solution. Since water is flowing in and out at the same rate, the volume V is constant. Next consider the amount y of pollutant in the basin as a function of time. Pollutant is flowing in the basin at a rate of

$$(r \text{ gallons/year})(c_1 e^{-\alpha t} \text{ pounds/gallon}) = r c_1 e^{-\alpha t} \text{ pounds/year,}$$

and out of the basin at a rate of

$$(r \text{ gallons/year})(\frac{y}{V} \text{ pounds/gallon}) = \frac{r}{V} y \text{ pounds/year.}$$

Therefore the amount of pollutant y satisfies the differential equation

$$\frac{dy}{dt} = r c_1 e^{-\alpha t} - \frac{r}{V} y.$$

This differential equation is not separable, because there is no way of writing the right-hand side as a function of t times a function of y. It is, however, first order linear:

$$y' + \frac{r}{V} y = r c_1 e^{-\alpha t}.$$

Hence if $\rho = e^{\int \frac{r}{V} dt} = e^{\frac{r}{V} t}$,

$$
\begin{aligned}
y &= \rho^{-1} \left(\int \rho r c_1 e^{-\alpha t} dt + C \right) \\
&= e^{-\frac{r}{V} t} \left(\int r C_1 e^{(\frac{r}{V} - \alpha)t} dt + C \right) \\
&= e^{-\frac{r}{V} t} \left(\frac{r c_1}{\frac{r}{V} - \alpha} e^{(\frac{r}{V} - \alpha)t} + C \right) \\
&= \frac{r c_1}{\frac{r}{V} - \alpha} e^{-\alpha t} + C e^{-\frac{r}{V} t}.
\end{aligned}
$$

Therefore,

$$c(t) = \frac{y}{V} = \frac{c_1 \frac{r}{V}}{\frac{r}{V} - \alpha} e^{-\alpha t} + C e^{-\frac{r}{V} t} \quad (\text{new } C).$$

There are two terms to the answer. The first dies off at the same rate as the entering pollution. The second dies off at a rate depending on how fast water is flowing in and out of the basin.

Part B. Check the answer for the special case $c_1 = 0$, which means that pollution stops altogether at time $t = 0$.

Solution. If $c_1 = 0$, the differential equation becomes the familiar

$$y' = -\frac{r}{V} y$$

for a quantity growing at a (negative) rate proportional to the amount, so the solution is

$$y = y_0 e^{-\frac{r}{V}t}, \quad c = \frac{y}{V} = c_0 e^{-\frac{r}{V}t},$$

which agrees with the previous answer when $c_1 = 0$.

Part C. If the pollution stops at time $t = 0$ as in Part B, how long will it take for the pollution concentration c to decrease to 10% of its initial value?

Solution. For the pollution to decrease to 10% of its initial value means that

$$e^{-\frac{r}{V}t} = .1, \quad -\frac{r}{V}t = \ln(.1) = -\ln(10), \quad t = \frac{V}{r}\ln(10).$$

For Lake Superior, it would take about 430 years.

29.6. The Famous Snowplow Problem

The following problem has developed a certain notoriety over the years. It goes like this:

Snow has been falling steadily for a little while when the snowplow starts out at noon. It plows a constant volume of snow per unit time. It covers two miles in the first two hours, but only one mile in the second two hours. When did the snow start?

Solution. Let t be the number of hours past noon. The snow started at $t = -a$. Our goal is to find a.

Let $y(t)$ be the number of miles plowed by time t. Since the plow handles a constant volume per unit time, the speed $\dot{y}$ decreases as the depth increases, and

$$\dot{y} = \frac{C_1}{\text{depth}}.$$

Since the snow falls steadily, the depth equals $C_2(t + a)$. Therefore

$$\frac{dy}{dt} = \frac{C_1}{C_2(t + a)},$$

and

$$y = C_3 \ln(t + a) + C_4.$$

Since $y(0) = 0$, $C_4 = -C_3 \ln a$, so that

$$y = C_3 \ln \frac{t + a}{a}.$$

Since $y(2) = 2$ and $y(4) = 3$,

$$2 = C_3 \ln \frac{2 + a}{a},$$

$$3 = C_3 \ln \frac{4 + a}{a}.$$

Dividing the first equation by the second yields

$$\frac{2}{3} = \frac{\ln \frac{2+a}{a}}{\ln \frac{4+a}{a}},$$

$$2 \ln \frac{4 + a}{a} = 3 \ln \frac{2 + a}{a}$$

$$\ln \left(\frac{4 + a}{a} \right)^2 = \ln \left(\frac{2 + a}{a} \right)^3$$

$$a(a^2 + 8a + 16) = a^3 + 6a^2 + 12a + 8$$

$$a^2 + 2a - 4 = 0$$

$$a = \frac{-2 \pm \sqrt{4 + 16}}{2} = -1 \pm \sqrt{5}.$$

Since $a > 0$, $a = \sqrt{5} - 1$. The snow started $\sqrt{5} - 1$ hours before noon, i.e., about 10:46 am.

Exercises 29

Solve the following differential equations.

1. $y' = \dfrac{\cos x}{y}$

2. $y' = \dfrac{x^2}{e^y}$

3. $y' = (1 + x^2)(1 + y^2)$

4. $y' = x\sqrt{1 - y^2}$

5. $y' + \dfrac{2}{x}y = x^3$

6. $y' + x^2 y = x^2$

7. $y' = \dfrac{x}{y}$

8. $y' = \dfrac{y}{x}$

9. $y' = \dfrac{y}{x} + x$

10. $y' = \dfrac{y}{x} + y$

11. $y' = \dfrac{x^2 - 1}{y + 1}$

12. $y' = \dfrac{y + 1}{x^2 - 1}$

13. $y' = \dfrac{2y + x}{x^2 + 2x}$

14. $y' = e^{x+y}$

15. $5y' + 3y + 2x = 0$

16. $5y' + xy + 2x = 0$

Solve the following initial value problems.

17. $y' = \dfrac{x^2}{y^3}, \quad y(1) = 1$

18. $y' = x + y, \quad y(0) = 1$

19. $2y' + 7y + 4x^2 = 0, \quad y(0) = 0$

20. $y' + 11y + \cos x = 0, \quad y\left(\dfrac{\pi}{2}\right) = 0$

Realistic problems often use other letters instead of x and y, but the methods are exactly the same. Solve the following differential equations.

21. $\dfrac{dA}{dz} = \dfrac{1}{2}A\sqrt{z}$

22. $\dfrac{dh}{dm} = a\,e^{-bm} - ch \quad$ (a, b, c constants)

23. $f' = \dfrac{1}{z}f - .01z^2$

24. $\dfrac{dr}{dt} = 12r - 100 \quad r(0) = r_0$

(This might represent the population of a rabbit farm supposed to produce continuously at the rate of 100 rabbits per year. How big must r_0 be for the farm to meet the demand?)

25. The moth balls protecting my Scottish wool sweater were 1 cm in radius when I packed it away May 1, but by June 1 they had shrunk to .5 cm. Assuming that they evaporate at a rate proportional to surface area, find the radius as a function of the number of months t since May 1. When will the moths get my sweater?

26. Derive the law of exponential population growth (see Sections 9.3 and 10.1)
$$p = p_0 e^{at},$$
from the differential equation
$$\frac{dp}{dt} = ap.$$

27. *Verhulst Equation.* a. Find the general solution $p(t)$ to the Verhulst Equation for population growth:
$$\frac{dp}{dt} = ap - bp^2 \quad (a, b > 0).$$

(The new $-bp^2$ term reflects crowding, pollution, etc., due to very large population. If you get stuck, do easier cases first, e.g., $b = 0$ or $a = b = 1$.)

b. Before checking against the answer, at least make sure your answer is reasonable. For $b = 0$, does it give the familiar solution $p = p_0 e^{at}$? What is
$$p_\infty = \lim_{t \to \infty} p(t)?$$

Use the differential equation to check that when $p = p_\infty$, $dp/dt = 0$.

c. Verhulst studied US population from 1790 to 1840. Let t be the number of years since 1790, when the population was 3.9 million. He fitted the best values $a = .03134$, $b = 1.5887 \times 10^{-10}$. What could he predict for population in 1870? 1930? 1990? As $t \to \infty$? (Observed values in millions are 38.6, 123, 246).

28. *Falling Objects.*

   a. Assume a tomato in frictionless free fall has a downward velocity v increasing at a constant rate g (the gravitational constant). Come up with a general solution for $v(t)$. (Let $v(0) = v_0$ to agree with most introductory physics texts.)

   b. Now add to the differential equation a negative air friction term proportional to the velocity $(-kv)$. Again obtain a general solution for $v(t)$. Show that the velocity approaches the constant g/k, called the terminal velocity.

  * c. Finally consider adding a second negative air friction term proportional to the square of the velocity $(-cv^2)$. For computational simplicity, let $g = 10$, $k = 3$, and $c = 1$. Solve for $v(t)$. Show that the terminal velocity is 2.

29. Consider the differential equation

$$y' = \frac{x^2 - y^2}{x^2 + y^2}.$$

For which values of x and y is y increasing?

Mixing.

30 Work the example of Section 29.5 yourself with the book closed.

31 Initially, Lake Euler has pollutant concentration k grams/gallon and volume V_0 gallons. Water with pollutant concentration of k grams/gallon flows in at the constant rate of r gallons/year. Water is drained out at the constant rate of $r/2$ gallons/year, and (pure) water evaporates at the constant rate of $r/2$ gallons/year.

   a. Solve for the pollutant concentration $c(t)$. Check that as $t \to \infty$, $c(t) \to 2k$.

  * b. Solve for the pollutant concentration $c(t)$ for the case of no evaporation, initial pollutant concentration 0 (other rates unchanged). Check that as $t \to \infty$, $c(t) \to k$. Hint: the volume $V = V_0 + \frac{r}{2}t$.

   c. Same as Part b except with initial pollutant concentration k.

30

Linear Second Order Homogeneous Constant-Coefficient Differential Equations

The long title of this section just refers to the important differential equations of the form

$$y'' + a_1 y' + a_0 y = 0. \tag{30.1}$$

It is *linear* in y, with no $(y')^2$ or e^y, etc. It is *second order* because it involves the second derivative. It is *homogeneous* because the right-hand side is 0 instead of some general function of x. The *coefficients* a_1, a_0 are *constants*, not functions of x.

To solve, just write down the *characteristic equation* which results from replacing the second derivative y'' by r^2, the first derivative y' by r, and the "zeroth derivative" y by $r^0 = 1$:

$$r^2 + a_1 r + a_0 = 0;$$

and solve for the roots r_1, r_2. The general solution to the original differential equation is then given by the following table.

Table 30.1. The roots of the characteristic equation determine the form of the general solution.

Roots r_1, r_2 of Char. Eq.	General Solution of Diff. Eq.
r_1, r_2 real and unequal	$C_1 e^{r_1 x} + C_2 e^{r_2 x}$
Double real root	$C_1 e^{rx} + C_2 x e^{rx}$
Complex roots $a \pm ib$	$e^{ax}(C_1 \cos bx + C_2 \sin bx)$

Notice that for these *second order* differential equations, there are two constants of integration. To solve for the constants, you need two additional bits of data, such as the initial values of y and y' when $x = 0$ or $x = a$.

Example 30.1. Solve the initial value problem

$$y'' + 4y' + 6y = 0, \quad y(0) = 1, \ y'(0) = 0.$$

The characteristic equation is $r^2 + 4r + 6 = 0$. First try to factor it. This one looks like it should factor, but it does not. Finally resort to the quadratic formula:

$$r = \frac{-4 \pm \sqrt{16 - 24}}{2} = -2 \pm i\sqrt{2}.$$

The roots are complex $a \pm ib$, with $a = -2$ and $b = \sqrt{2}$. By Table 30.1, the general solution is:

$$y = e^{-2x}\left(C_1 \cos \sqrt{2}x + C_2 \sin \sqrt{2}x\right).$$

To plug in the initial value to solve for C_1 and C_2, we first need to differentiate the solution to get a formula for y':

$$
\begin{aligned}
y' &= -2e^{-2x}\left(C_1 \cos \sqrt{2}x + C_2 \sin \sqrt{2}x\right) \\
&\quad + e^{-2x}\left(-\sqrt{2}C_1 \sin \sqrt{2}x + \sqrt{2}C_2 \cos \sqrt{2}x\right) \\
&= e^{-2x}\left(\left(-2C_1 + \sqrt{2}C_2\right)\cos \sqrt{2}x + \left(-2C_2 - \sqrt{2}C_1\right)\sin \sqrt{2}x\right).
\end{aligned}
$$

Plugging the initial values $y(0) = 1$ and $y'(0) = 0$ into the equations for y and y' yields

$$\begin{aligned} 1 &= C_1 \\ 0 &= -2C_1 + \sqrt{2}C_2. \end{aligned}$$

Now we solve these two equations in the two unknowns. This set is easy — it can be much worse. Here $C_1 = 1$. Hence

$$0 = -2 + \sqrt{2}C_2 \Rightarrow C_2 = \sqrt{2}.$$

The solution is

$$y = e^{-2x}\left(\cos\sqrt{2}x + \sqrt{2}\sin\sqrt{2}x\right).$$

Derivation of Table. Since the solution to the *first* order linear homogeneous constant-coefficient differential equation $y' + a_0 y = 0$, i.e., $y' = -a_0 y$, is $y_0 e^{a_0 x}$, it is reasonable to look for solutions of the form $y = Ce^{rx}$. Plugging that into the original Equation 30.1 yields

$$r^2 Ce^{rx} + a_1 r Ce^{rx} + a_0 Ce^{rx} = 0.$$

Canceling Ce^{rx} (e^{rx} is never 0) yields

$$r^2 + a_1 r + a_0 = 0.$$

There is the characteristic equation! This is where it comes from. Hence the roots yield solutions $C_1 e^{r_1 x}$ and $C_2 e^{r_2 x}$. It is easy to check that sums $C_1 e^{r_1 x} + C_2 e^{r_2 x}$ are also solutions. If the roots are distinct, these turn out to be all the solutions. (If r_1, r_2 are complex, the complex exponentials must be interpreted as sines and cosines, using the amazing identity $e^{ibx} = \cos bx + i\sin bx$.) If the roots are equal, a solution is missing. It turns out to be Cxe^{rx}.

30.1. Damped Vibrations

A mass m on a spring distorted from the rest position $x = 0$ feels a restoring force $-kx$ proportional to the displacement from rest and

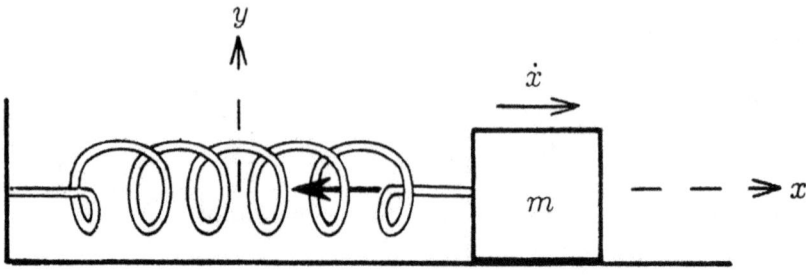

$$F = -kx - c\dot{x}$$

Figure 30.1. A mass on a spring feels a restoring force $-kx$ and a damping force $-c\dot{x}$. Thus the total force $F = -c\dot{x} - kx$.

opposite in direction. See Figure 30.1. Friction in air or some liquid is roughly proportional to the speed $\dot{x}$ but opposite in direction: $-c\dot{x}$. Thus the total force F is $-c\dot{x} - kx$.

By Newton's Law, $F = ma = m\ddot{x}$. Therefore

$$m\ddot{x} + c\dot{x} + kx = 0.$$

This is a linear second order constant-coefficient differential equation for $x(t)$ instead of $y(x)$. The general behavior will depend on the roots of the characteristic equation

$$mr^2 + cr + k = 0,$$

$$r = \frac{-c \pm \sqrt{c^2 - 4km}}{2m}.$$

If the damping constant c is relatively large so that the discriminant

$$D = c^2 - 4km > 0,$$

the roots will be real, distinct (and negative); the solution will be

$$x = c_1 e^{r_1 t} + c_2 e^{r_2 t} \qquad (r_1, r_2 < 0);$$

Table 30.2. The behavior of a vibrating mass, depending on the spring constant k and the damping constant c.

| Damped Vibrations | $m\ddot{x} + c\dot{x} + kx = 0$ |

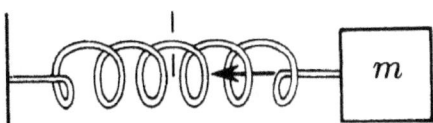

Spring Constant k

Damping Constant c

1. Overdamped if $c^2 > 4km$

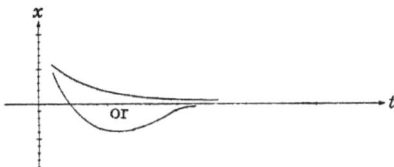

2. Critically damped if $c^2 = 4km$

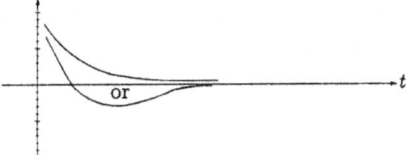

3. Underdamped if $0 < c^2 < 4km$

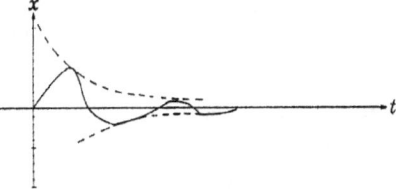

4. Undamped if $c = 0$

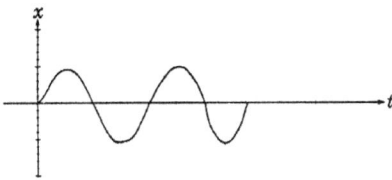

and the mass will head exponentially back to $x = 0$. If the damping constant c is relatively small so that the discriminant

$$D = c^2 - 4km \ < \ 0,$$

the roots will be complex

$$r = -\frac{c}{2m} \pm ib, \quad \text{where} \quad b = \frac{1}{2m}\sqrt{|D|},$$

the solution will be

$$x = e^{-\frac{c}{2m}t}(c_1 \cos bt + c_2 \sin bt),$$

and the mass will oscillate with amplitude $\approx e^{-\frac{c}{2m}t}$, which approaches 0 as t grows large. If the damping constant $c = 0$, the mass will simply oscillate.

In the borderline case $D = 0$, $\frac{c}{2m}$ is a double root, and the solution will be

$$x = e^{-\frac{c}{2}t}(c_1 + c_2 t),$$

and the mass heads exponentially back to $x = 0$. Actually it is in this borderline case that the mass approaches 0 soonest.

The various types of damped vibrations are summarized in Table 30.2.

Example 30.2. Predict $\lim_{x \to \infty} y$ for a solution y of

$$y'' + (3 + \cos x)y' + 5y = 0.$$

Solution. We have not studied such an equation with variable coefficients, but we expect it to behave like a mass or a spring with variable positive damping. Hence we predict $y \to 0$.

Exercises 30

Find the general solution.

1. $y'' - 5y' + 6y = 0$

2. $y'' + 2y' - 15y = 0$

3. $y'' - 6y' + 9y = 0$

4. $y'' + 4y' + 5y = 0$

5. $y'' + 9y = 0$

6. $y'' + 2y' + 3y = 0$

7. $3y'' + 2y' + y = 0$

8. $y'' + 9y' = 0$

Solve the initial value problem.

9. $y'' - 10y' + 21y = 0$ $\qquad\qquad$ $y(0) = y'(0) = 0$

10. $y'' + 6y' - 16y = 0$ $\qquad\qquad$ $y(0) = 1, y'(0) = 0$

11. $y'' - 8y' + 15y = 0$ $\qquad\qquad$ $y(0) = 2, y'(0) = -3$

* 12. $y'' + 4y + 2 = 0$ $\qquad\qquad$ $y(0) = -1/2, y'(0) = 0$

13. Predict the behavior as $x \to \infty$ of $5y'' + xy' + 3y = 0$.

14. Give the solution or at least the method if any.
   a. $x^2 y' + y^2 + 1 = 0$
   b. $x^2 y' + y + x = 0$
   c. $x^2 y''' + 5x^2 = 0$

15. When do you use the formula $y' = r_{in} c_{in} - r_{out} c_{out}$?

Part V

Multivariable Calculus

31

Partial Derivatives

A function of two variables such as

$$z = f(x, y) = ye^{xy}$$

changes if x changes and changes if y changes. There are two parts to the derivative, called partial derivatives:

$$\frac{\partial f}{\partial x} = \text{rate of change of } f \text{ as } x \text{ changes, } y \text{ constant,}$$

$$\frac{\partial f}{\partial y} = \text{rate of change of } f \text{ as } y \text{ changes, } x \text{ constant,}$$

(pronounced "del f del x" and "del f del y"). The funny symbol ∂ called "del" is used instead of d to remind you that each partial derivative is just part of the derivative. Instead of $\frac{\partial f}{\partial x}$ you can write f_x ("f sub x") or $\frac{\partial z}{\partial x}$ "del z del x") or z_x ("z sub x").

Fortunately $\frac{\partial f}{\partial x}$ is easy to compute: you just treat y as a constant. Since the ordinary derivative of ae^{ax} is a^2e^{ax}, if

$$f(x, y) = ye^{xy}, \quad \frac{\partial f}{\partial x} = y^2e^{xy}.$$

Similarly, to compute $\frac{\partial f}{\partial y}$, you treat x as a constant. For our example, $\frac{\partial f}{\partial y}$ is a bit harder, because y occurs twice and we have to use the product rule:

$$\frac{\partial}{\partial y}(ye^{xy}) = \frac{\partial}{\partial y}(y) \cdot e^{xy} + y \cdot \frac{\partial}{\partial y}(e^{xy})$$

$$= 1 \cdot e^{xy} + y \cdot xe^{xy} = (1 + xy)e^{xy}.$$

Second derivatives. You can take second derivatives (derivatives of derivatives), but now there are four possibilities. You can start with f_x and take its partial derivative with respect to x or y to get f_{xx} and f_{xy} (pronounced "$f\ x\ x$," "$f\ x\ y$"). Or you can start with f_y and take its partial derivatives f_{yx} and f_{yy}. For our example:

$$f_{xx} = \frac{\partial}{\partial x}(f_x) = \frac{\partial}{\partial x}(y^2 e^{xy}) = y^3 e^{xy},$$

$$f_{xy} = \frac{\partial}{\partial y}(f_x) = \frac{\partial}{\partial y}(y^2 e^{xy}) = 2ye^{xy} + y^2 xe^{xy}$$

$$= (2y + xy^2)e^{xy},$$

$$f_{yx} = \frac{\partial}{\partial x}(f_y) = \frac{\partial}{\partial x}((1 + xy)e^{xy}) = ye^{xy} + (1 + xy)ye^{xy}$$

$$= (2y + xy^2)e^{xy},$$

$$f_{yy} = \frac{\partial}{\partial y}(f_y) = \frac{\partial}{\partial y}((1 + xy)e^{xy}) = xe^{xy} + (1 + xy)xe^{xy}$$

$$= (2x + x^2 y)e^{xy}.$$

Do you think it is a coincidence that $f_{yx} = f_{xy}$? It is not. It is generally true that $f_{yx} = f_{xy}$, so we do not have to worry about the order of partial differentiation, and there are really just three distinct second partial derivatives: f_{xx}, f_{xy}, and f_{yy}. These can also be written $\frac{\partial^2 f}{\partial x^2}, \frac{\partial^2 f}{\partial x \partial y}$, and $\frac{\partial^2 f}{\partial y^2}$ (pronounced "del squared f del x squared," "del squared f del x del y," and "del squared f del y squared.")

If $z = f(x, y)$, the various second partial derviatives may also be written

$$z_{xx}, z_{xy}, z_{yy} \quad \text{or} \quad \frac{\partial^2 z}{\partial x^2}, \frac{\partial^2 z}{\partial x \partial y}, \frac{\partial^2 z}{\partial y^2}.$$

Example 31.1. Suppose $z = xy \ln x$. Find all second partials.

Solution. By the product rule for $z = (xy)(\ln x)$,

$$z_x = y \ln x + (xy)(1/x) = y \ln x + y, \quad z_y = x \ln x,$$

$$z_{xx} = y/x, \quad z_{xy} = z_{yx} = \ln x + 1, \quad z_{yy} = 0.$$

Exercises 31

Find the partial derivatives $f_x = \dfrac{\partial f}{\partial x}$ and $f_y = \dfrac{\partial f}{\partial y}$.

1. $f(x, y) = x^2 y + xy^3$

2. $f(x, y) = x^3 + 3x^2 y + 3xy^2 + y^3$

3. $f(x, y) = (x + y)^3$

4. $f(x, y) = e^{-x^2 - y^2}$

5. $f(x, y) = e^{y/x}$

6. $f(x, y) = xye^{xy}$

7. $f(x, y) = \frac{x^2 - y^2}{x^2 + y^2}$

8. $f(x, y) = x \ln(x + y)$

9. $f(x, y) = 3x \cos 2y$

10. $f(x, y) = \tan xy^2$

11. If $f(x, y, z) = x^2y^2z^2 + 2x^2 + 3y^3 + 4z^4$, find f_x, f_y, and f_z.

12. If $f(r, h) = \pi r^2 h$, find $\dfrac{\partial f}{\partial r}$ and $\dfrac{\partial f}{\partial h}$.

13. According to the ideal gas law, the volume V satisfies $V = nRT/p$, where the number of moles n and R are constants and the temperature T and pressure p are variables. Find $\partial V/\partial T$ and $\partial V/\partial p$.

14. If $f(a, b, \theta) = ab \sin \theta$, find $\dfrac{\partial f}{\partial a}$, $\dfrac{\partial f}{\partial b}$, and $\dfrac{\partial f}{\partial \theta}$.

15. If $f(a, b, \theta) = a^2 + b^2 - 2ab \cos \theta$, find $\dfrac{\partial f}{\partial a}$, $\dfrac{\partial f}{\partial b}$, and $\dfrac{\partial f}{\partial \theta}$.

Find the second partials z_{xx}, $z_{xy} = z_{yx}$, and z_{yy}.

16. $z = x^3 + 5xy + y^3$

17. $z = x^4y^2$

18. $z = x^4 + 4x^3y + 6x^2y^2 + 4xy^3 + y^4$

19. $z = (x^2 + y^2)^{10}$

20. $z = e^{-\frac{x^2+y^2}{2}}$

21. $z = \ln x^3y^5$

22. $z = \tan^{-1} \frac{y}{x}$

32

Double Integrals

Just as a function $f(x, y)$ of two variables may be differentiated with respect to x or y (holding the other constant), it may be integrated with respect to x and y (each time holding the other constant). To evaluate a double integral such as

$$\int_{x=0}^{2} \underbrace{\int_{y=0}^{x} e^{y/x} \, dy}_{\text{INSIDE}} \, dx,$$

$$\underbrace{\hspace{5cm}}_{\text{OUTSIDE}}$$

you start with the inside integral (with respect to y in this case), treating the outside variable x as constant. Since $\int e^{ay} \, dy = \frac{1}{a}e^{ay}$, taking $a = 1/x$ the inside integral becomes

$$\text{INSIDE} = xe^{y/x}\Big]_{y=0}^{x} = xe^1 - xe^0 = (e-1)x.$$

Now we integrate that answer with respect to the outside variable x:

$$\text{OUTSIDE} = (e-1)\frac{1}{2}x^2\Big]_0^2 = (e-1)(2-0) = 2(e-1) \approx 3.44.$$

Notice that the limits of integration on the inside integral (here 0 and x) can depend on the outside variable, while the limits of integration on the outside integral (here 0 and 2) must be constants.

Exercises 32

1. $\displaystyle\int_{x=1}^{2}\int_{y=2}^{x} x^2 y \, dy \, dx$

2. $\displaystyle\int_{x=2}^{3}\int_{y=x}^{x^2} y \, dy \, dx$

3. $\displaystyle\int_{x=-2}^{2}\int_{y=0}^{x} (x+y)^3 \, dy \, dx$

4. $\displaystyle\int_{x=0}^{5}\int_{y=0}^{3} 2 \, dy \, dx$

5. $\displaystyle\int_{x=2}^{3}\int_{y=\ln x}^{2\ln x} e^y \, dy \, dx$

6. $\displaystyle\int_{y=0}^{2}\int_{x=0}^{y} xy e^{x^2+y^2} \, dx \, dy$

7. $\displaystyle\int_{0}^{\pi/2}\int_{0}^{\pi/2} x^2 \cos y \, dy \, dx$

8. $\displaystyle\int_0^{10}\int_0^{\pi/4} \sin^2 x \, \cos x \, dx \, dy$

9. $\displaystyle\int_0^1\int_0^1 \frac{x^2}{1+y^2} \, dy \, dx$

10. $\displaystyle\int_0^2\int_0^1 \frac{x^2}{\sqrt{4-y^2}} \, dy \, dx$

11. $\displaystyle\int_0^1\int_0^h \frac{4}{3}\pi r^3 h \, dr \, dh$

12. $\displaystyle\int_1^{12}\int_0^{3/V} pV \, dp \, dV$

13. $\displaystyle\int_0^1\int_0^x\int_0^y z \, dz \, dy \, dx$

33

Critical Points

Recall that in finding maxima and minima of a function $f(x)$ of one variable x, it was important to find the critical points where the derivative $f'(x) = 0$. Similarly in finding maxima and minima of a function $f(x, y)$ of two variables, it will be important to find the critical points where both partial derivatives f_x and f_y are 0.

Example 33.1. Find the critical points of

$$f(x, y) = x^2 + 4xy + y^3.$$

Solution. Set the partial derivatives equal to 0:

$$0 = f_x = 2x + 4y$$

$$0 = f_y = 4x + 3y^2$$

Solving for $x = -2y$ in the first equation and substituting in the second equation yields

$$0 = -8y + 3y^2 = y(-8 + 3y)$$

So $y = 0$ or $y = 8/3$. Since $x = -2y$, the critical points are $(0, 0)$ and $(-16/3, 8/3)$.

Exercises 33

Find all critical points

1. $f(x, y) = x^2 - 2x + y^3 - 12y$

2. $f(x, y) = \frac{1}{3}x^3 - 3x^2 + 5x + y^3 - 3y + 8$

3. $f(x, y) = x^2 + 6xy + y^3$

4. $f(x, y) = x^2 + 2x + 6xy + y^3 + 6y$

5. $f(x, y) = x^4 + y^4 - 4xy$

6. $f(x, y) = x^4 + 16y^4 - 8xy$

7. $f(x, y) = (x - y)^2$

34

Maxima and Minima

Just as with a function $f(x)$ of one variable, to find the absolute maxima or minima of a function $f(x, y)$ of two variables, look at

1. *extreme cases* (sometimes given, sometimes x or $y \to \pm\infty$) or f undefined,

2. *critical cases*, where $f_x = f_y = 0$ (or f_x or f_y is undefined).

Example 34.1. Find the absolute maxima and minima of the function

$$f(x, y) = x^2 + y^2 - x$$

on the strip $|x| \leq 1$ of Figure 34.1.

Solution. As $y \to \pm\infty$, $f(x, y) \to +\infty$, so the function never reaches a maximum. Since $|x| \leq 1$, we do not consider $x \to \pm\infty$, but we do need to check the boundary for possible minima (the only reason we do not need to check for maxima is that we already know f has no absolute maxima). On the left boundary, $x = -1$ and $f(x, y) = (-1)^2 + y^2 - (-1) = y^2 + 2$, which hits its smallest value

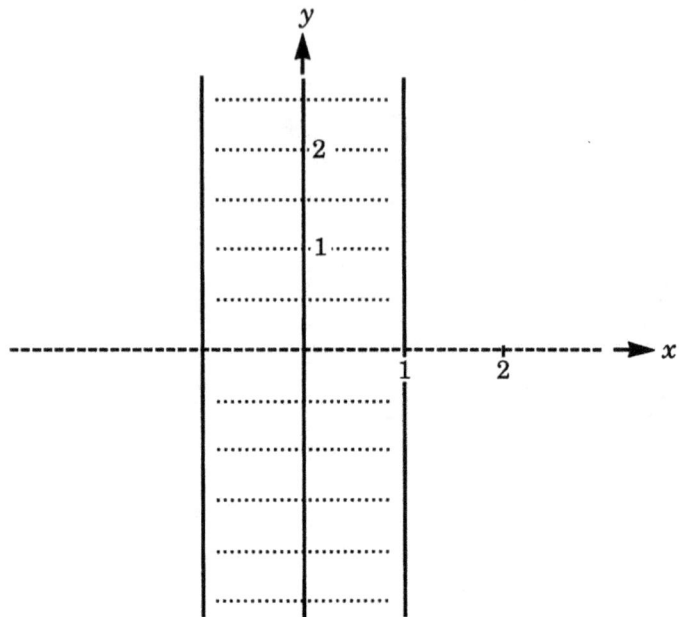

Figure 34.1. The pictured strip $|x| \leq 1$ is the domain on which we want to find the maxima and minima of $f(x, y) = x^2 + y^2 - x$.

of 2 when $y = 0 : f(-1, 0) = 2$. This is our first candidate for an absolute minimum. On the right boundary, $x = 1$ and $f(x, y) = y^2$, which hits its smallest value of 0 when $y = 0 : f(1, 0) = 0$. This is our second and better candidate for an absolute minimum.
Second we check critical points:

$$0 = f_x = 2x - 1 \qquad \Rightarrow x = 1/2$$

$$0 = f_y = 2y \qquad \Rightarrow y = 0$$

There is just one critical point $(1/2, 0)$, where $f(1/2, 0) = \frac{1}{4} + 0 - \frac{1}{2} = -\frac{1}{4}$. This is our third candidate for a minimum. Comparing our three candidates we see that the last is the absolute minimum:

$$f(1/2, 0) = -1/4.$$

Exercises 34

Find the absolute maxima and absolute minima.

1. $f(x, y) = x^2 + y^2$

2. $f(x, y) = x^4 + x^2 y^2 + y^4$

3. $f(x, y) = x^2 + y^2$ for $1 \leq x \leq 2$

4. $f(x, y) = x^2 - 2x + y^2$ for $-1 \leq x \leq 2$

5. $f(x, y) = x^2 + y^2$ for $-1 \leq x \leq 2, -1 \leq y \leq 3$

6. $f(x, y) = x^2 - 2x + y^2$ for $-1 \leq x \leq 2, -1 \leq y \leq 3$

7. The temperature T in Captain Jones's swimming pool ($-1 \leq x \leq 1, -2 \leq y \leq 3$) on Venus is given by the function $T(x, y) = 100 - 4x^2 - y^2 + x^2 y^2$. Captain Jones likes to keep cool; his partner Artudeetu favors the hotter temperatures. Where is Captain Jones's favorite spot? Artudeetu's?

8. Find the maxima and minima of

$$f(x, y) = 4x^2 + \frac{3}{4}y^2 + xy^2 - x^2 y^2 - 4x$$

for $-1 \leq x \leq 2, -1 \leq y \leq 3$.

Answers to Odd-Numbered Exercises

Chapter 1

1. $7t^6 + 5t^4$

3. $-18t^2 + 24t - 4$

5. $t^9 + t^8 + t^7$

7. $18t + 30$, because $f(t) = 9t^2 + 30t + 25$

9. $-48t^5 + 48t^3 - 12t$,
   because $f(t) = -8t^6 + 12t^4 - 6t^2 + 1$

11. $12t^3 - 12t^2 - 18t + 8$, because $f(t) = 3t^4 - 4t^3 - 9t^2 + 8t - 1$

13. $f'(t) = \lim_{\Delta t \to 0} \dfrac{f(t + \Delta t) - f(t)}{\Delta t} = \lim_{\Delta t \to 0} \dfrac{(3(t + \Delta t) + 5) - (3t + 5)}{\Delta t}$

$\quad = \lim_{\Delta t \to 0} \dfrac{3\Delta t}{\Delta t} = \lim_{\Delta t \to 0} (3) = 3$

15. $f'(t) = \lim_{\Delta t \to 0} \dfrac{f(t + \Delta t) - f(t)}{\Delta t}$

$\quad = \lim_{\Delta t \to 0} \dfrac{((t + \Delta t)^2 + (t + \Delta t) + 1) - (t^2 + t + 1)}{\Delta t}$

$\quad = \lim_{\Delta t \to 0} \dfrac{t^2 + 2t\Delta t + \Delta t^2 + t + \Delta t + 1 - t^2 - t - 1}{\Delta t}$

$\quad = \lim_{\Delta t \to 0} \dfrac{2t\Delta t + \Delta t^2 + \Delta t}{\Delta t} = \lim_{\Delta t \to 0} (2t + \Delta t + 1) = 2t + 1$

17. $10x + 7$

19. $y' = \lim\limits_{\Delta x \to 0} \dfrac{y(x + \Delta x) - y(x)}{\Delta x} = \lim\limits_{\Delta x \to 0} \dfrac{\frac{1}{2}(x + \Delta x)^2 - \frac{1}{2}x^2}{\Delta x}$

$= \lim\limits_{\Delta x \to 0} \dfrac{\frac{1}{2}(x^2 + 2x\Delta x + \Delta x^2) - \frac{1}{2}x^2}{\Delta x}$

$= \lim\limits_{\Delta x \to 0} \dfrac{x\Delta x + \frac{1}{2}\Delta x^2}{\Delta x} = \lim\limits_{\Delta x \to 0}\left(x + \dfrac{1}{2}\Delta x\right) = x$

Chapter 2

1.

x	-3	-2	-1	0	1	2	3
$y = x^2 - 4$	5	0	-3	-4	-3	0	5
My slope guess	-6	-3	$-\frac{3}{2}$	0	$\frac{3}{2}$	3	6
Slope $y' = 2x$	-6	-4	-2	0	2	4	6

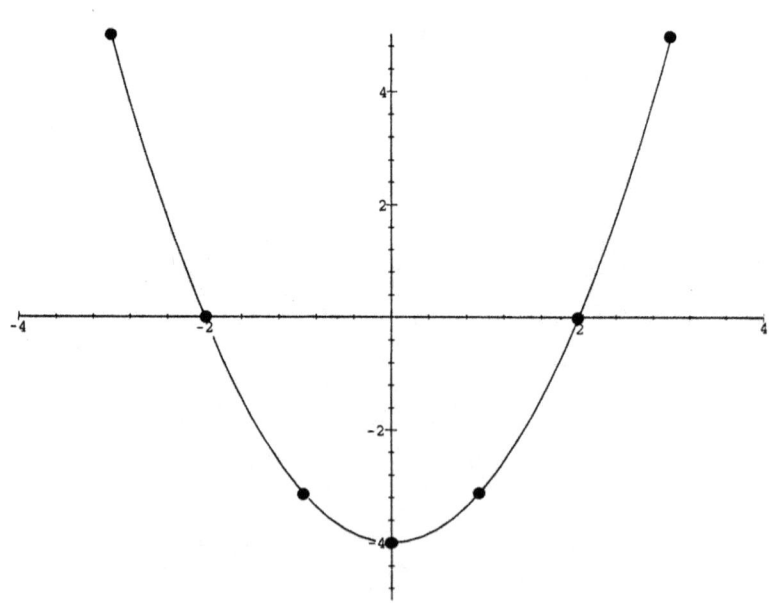

3.

x	-3	-2	-1	0	1	2	3
$y = -\frac{x^2}{3} + \frac{x}{3} + 2$	-2	0	$\frac{4}{3}$	2	2	$\frac{4}{3}$	0
My slope guess	2	$\frac{3}{2}$	1	$\frac{1}{3}$	$-\frac{1}{3}$	-1	$-\frac{3}{2}$
Slope $y' = -\frac{2}{3}x + \frac{1}{3}$	$\frac{7}{3}$	$\frac{5}{3}$	1	$\frac{1}{3}$	$-\frac{1}{3}$	-1	$-\frac{5}{3}$

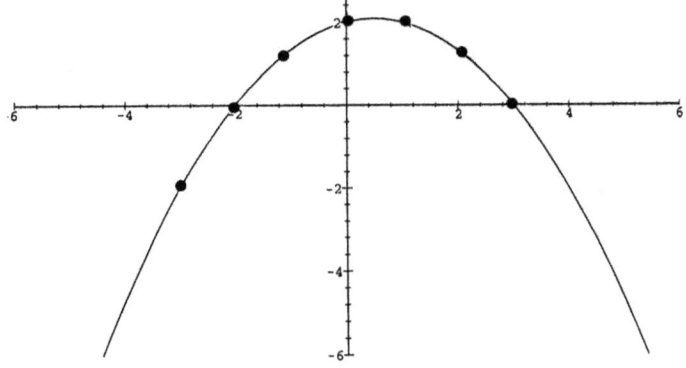

5.

x	-3	-2	-1	0	1	2	3
$y = x^3 - 9x$	0	10	8	0	-8	-10	0
My slope guess	15	0	-6	-8	-6	0	15
Slope $y' = 3x^2 - 9$	18	3	-6	-9	-6	3	18

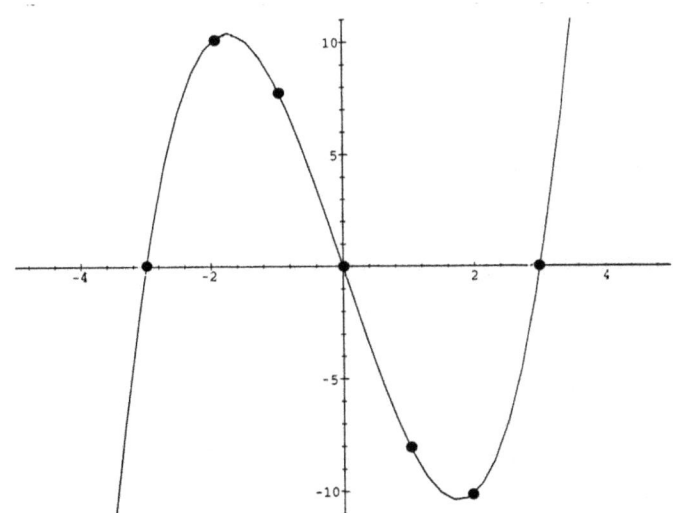

7. $y' = 2x, m = y'(2) = 4.$ Line $y = 4x + b$;
   Plugging in $(2,0) \rightarrow b = -8$ so $y = 4x - 8$

9. $y' = 5x^4$, $m = y'(1) = 5.$ Line $y = 5x + b = 5x - 4.$

Chapter 3

1. $(2x)(x^3 + 1) + (x^2 + 1)(3x^2) = 5x^4 + 3x^2 + 2x$

3. $6x^5$

5. $(-5x^2 - 12x + 5)/(x^2 + 1)^2$

7. $(-2x^7 + 4x^5 - 2x^3 + 3x^2 - 1)/(x^5 - x^3 + 1)^2$

9. x [No need to use quotient rule, although it does work.]

11. $6x^5 + 5x^4 + 4x^3 + 6x^2 + 2x + 1$

13. $(2x^7 - 3x^6 + 17x^4 - 52x^3 + 27x^2 - 24x + 20)/(x^3 + 1)^2$

15. $(-2x^{13} - 4x^{11} + x^{10} - 6x^9 + 2x^8 - 8x^7 - 2x^6 + 6x^5 - 6x^4 + 4x^3 - 3x^2 + 2x) \div (x^3 + 1)^2 (x^5 + 1)^2$

17. $f'(t) = (t^2 + t + 1)(\frac{1}{5}t^5 + \frac{1}{4}t^4 + t + 1) + (\frac{1}{3}t^3 + \frac{1}{2}t^2 + t + 1)(t^4 + t^3 + 1) = \frac{8}{15}t^7 + \frac{77}{60}t^6 + \frac{39}{20}t^5 + \frac{9}{4}t^4 + \frac{7}{3}t^3 + \frac{5}{2}t^2 + 3t + 2$

19. $f(t) = g(t)g(t);\ f'(t) = g'(t)g(t) + g(t)g'(t) = 2g(t)g'(t);$
    $f'(1) = 2g(1)g'(1)$
    Warning: You cannot plug in 1 first and then differentiate. The derivative of a constant is 0.

Chapter 4

1. $2(x^2 + 2x + 3)(2x + 2) = 4(x + 1)(x^2 + 2x + 3)$

3. $5x^2(\frac{x^3}{3} + 1)^4 + 4x(\frac{x^2}{2} + 1)^3$

5. $160x(x^2 + 1)^9((x^2 + 1)^{10} + 1)^7$

7. $[5\left((x^3 + 7)^4 + x\right)^4 \left(12x^2(x^3 + 7)^3 + 1\right)(x^2 + 1) - 2x\left((x^3 + 7)^4 + x\right)^5] \div (x^2 + 1)^2$

9. $-x^2/y^2$

11. $-x(2x^2 + y^2)/(x^2y + 2y^3)$

13. $\frac{3}{5}x^2/y^4$ $(= \frac{3}{5}x^{-2/5})$

15. $10x^9/(x+1)^{11}$

17. $6(3t+5)$

19. $-12t(-2t^2+1)^2$

21. $\frac{122}{169}$

Chapter 5

1. a. $8, 4, \frac{1}{2}$

   b. $27, \frac{1}{8}, \frac{27}{8}, \frac{8}{27}$

   c. x^3, cannot be simplified, $x^2, x^{-2} = \frac{1}{x^2}, x$

   d. $x^{\frac{4}{3}}, xy^2$, cannot be simplified, cannot be simplified

3. $\frac{3}{4}x^{-1/4} + \frac{4}{3}x^{1/3}$

5. $-\frac{1}{2}x^{-3/2}$

7. $\frac{2}{3}(2x-3)^{-2/3}$

9. $\frac{1}{2}(\sqrt{x+1}+2)^{-\frac{1}{2}}\frac{1}{2}(x+1)^{-\frac{1}{2}} = \dfrac{1}{4\sqrt{(\sqrt{x+1}+2)(x+1)}} = \dfrac{1}{4\sqrt{(x+1)^{3/2}+2x+2}}$

11. $y' = \frac{1}{2}(1+\sqrt{1+\sqrt{x}})^{-\frac{1}{2}}(\frac{1}{2}(1+\sqrt{x})^{-\frac{1}{2}})(\frac{1}{2}x^{-\frac{1}{2}}) = \dfrac{1}{8\sqrt{x+x\sqrt{x}+x(1+\sqrt{x})^{3/2}}}$

Chapter 6

1. $5\cos x$

3. $-20\cos 5x + 3\sin \frac{1}{2}x$; $y'(\pi/2) = 3/\sqrt{2}$

5. $-2x\sin x^2$

7. $-2\sin x \cos x = -\sin 2x$

9. $5\sin^4 x \cos x - 9\cos x$; $y'(3\pi/4) = -5/4\sqrt{2} + 9/\sqrt{2} = 31/4\sqrt{2}$

11. $-x(x^2+7)^{-1/2}\sin\sqrt{x^2+7}$

13. $-2/(1-2\sin x\cos x) = -2/(1-\sin 2x)$

15. $-3\sin 3x\cos 3x(\sin^2 3x + 7)^{-3/2}$

Chapter 7

1. Max -4 at 0, min $-6\frac{1}{4}$ at $1\frac{1}{2}$

3. Max 0 at 0 and 1, min $-\frac{4}{27}$ at $\left(\frac{2}{3}\right)^6$

5. No max, min $-13/4$ at $5/2$

7. No max, no min

9. a.   b.

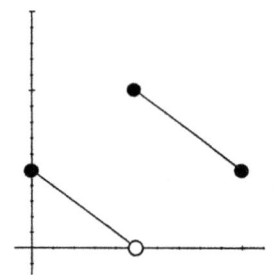
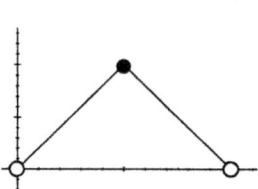

    c. not possible

11. a. $+\infty$ b. $+\infty$ c. $+\infty$

13. $3/4$

15. 0

17. $+\infty$

19. Min at $r = 10/\sqrt[3]{\pi} \approx 6.83$ of $2\pi\frac{100}{\pi^{2/3}} + \frac{4000}{10}\sqrt[3]{\pi} = 600\sqrt[3]{\pi} \approx 879$, no max

21. No max, min 0 at 0

23. a. No max, min at $x = \frac{-1+\sqrt{5}}{2} \approx .618$ of about -1.09
    b. No max, min 1 at $x = 0$

25. Max 1 at $2k\pi$, min -1 at $\pi + 2k\pi$

27. Max $\sqrt{2}$ at $3\pi/4 + 2k\pi$, min $-\sqrt{2}$ at $-\pi/4 + 2k\pi$

29. a. $-\infty$ b. $+\infty$ c. 0

31. a. $-\infty$ b. $+\infty$

33. a. $-\infty$ b. $+\infty$ c. $+\infty$ d. $-\infty$ e. 10 f. 10

35. No max, no min

37. Max ≈ 8.53 at $\frac{-3+\sqrt{65}}{4} \approx 1.266$, min $\approx .469$ at $\frac{-3-\sqrt{65}}{4} \approx -2.766$

39. Max 0 at 0, no min

41. Max 56 at 8 pm, min -25 at 5 pm

Chapter 8

1. $r = h = \sqrt[3]{V/\pi}$. Use r as variable; then $h = V/\pi r^2, A = \pi r^2 + 2\pi rh = \pi r^2 + 2Vr^{-1}$. Note that $a \to \infty$ as $r \to 0$ or ∞, so for minimum $A' = 0 \Rightarrow r = \sqrt[3]{V/\pi}$. $h = V/\pi r^2 = \frac{V}{\pi(V/\pi)^{2/3}} = \frac{V^{1/3}}{\pi^{1/3}} = \sqrt[3]{\frac{V}{\pi}}$.

3. 6 feet by 3 feet. Note extreme cases $x \to 0, h \to 0$ no good.

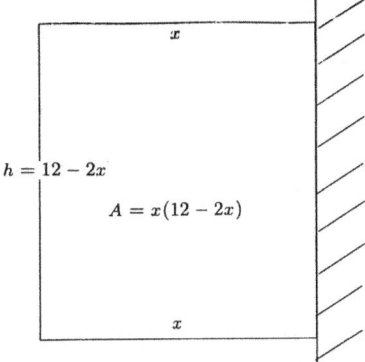

$h = 12 - 2x$

$A = x(12 - 2x)$

5. $a\sqrt{2} \times a/\sqrt{2}$, where a is the radius of the semicircle. $A = 2a^2 \sin\theta \cos\theta$.

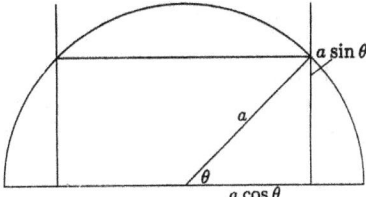

Note extreme cases no good:

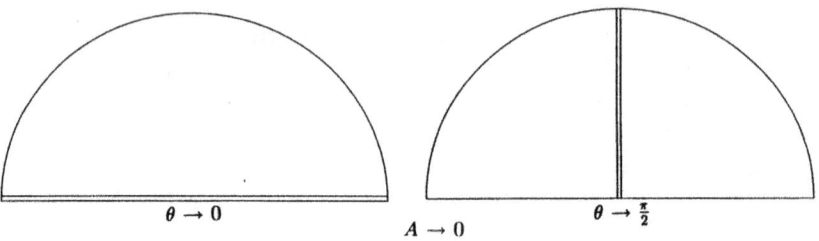

7. a. 250 feet by 250 feet   b. 250 feet by 500 feet

9. $4\sqrt{2}$ feet (same problem as 4!)

11. a. $A = 2x^2 + 4 \cdot x\frac{8000}{x^2} = 2x^2 + 32000x^{-1}$.

Ans: 20 in. by 20 in. by 20 in.

b. Cost $= 2x^2 + 2 \cdot x\frac{8000}{x^2} + 15 \cdot 2 \cdot x\frac{8000}{x^2} = 2x^2 + 256,000x^{-1}$.

Ans: 40 in. by 40 in. by 5 in.

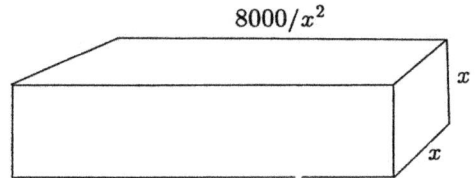

13. 20 times

15. 1/3

Chapter 9

1. 4

3. 5/3, because $8^{5/3} = (\sqrt[3]{8})^5 = 2^5 = 32$

5. -2, because $4^{-2} = 1/16$.

7. -2

9. 2

11. 4

13. $\sqrt{5}$ $(3^{\log_9 5} = 9^{\frac{1}{2}\log_9 5} = (9^{\log_9 5})^{\frac{1}{2}} = \sqrt{5})$

15. cannot be simplified

17. cannot be simplified

19. cannot be simplified

21. cannot be simplified

23. $5e^{5x}$

25. $\frac{1}{3}x^{-\frac{2}{3}}e^{\sqrt[3]{x}}$

27. $e^{\sin x}\cos x$

29. $e^{x\sin x}(\sin x + x\cos x)$

31. $(2x - 5)e^{\sin(x^2-5x-6)}\cos(x^2 - 5x - 6)$

33. e^{ex+1} because $(e^e)^x = e^{ex}$

35. $7^x \ln 7$

37. $2^{x^2+6x-5}(2x + 6)\ln 2$

39. $3x^2(\ln 5)5^{x^3}$

41. $-2(x - 1)^{-2}e^{\frac{x+1}{x-1}}$

43. $4x^3/(x^4 + 1)$

45. $e^x/(e^x + 1)$

47. $-2\cos x \sin x/(\cos^2 x + 1)$

49. $\frac{2}{x}\ln x$ $(= 2\ln x \cdot (\ln x)')$

51. $1/x$

53. $\frac{2x}{(\ln 10)(x^2+7)}$

55. $1/(x\ln x \ln\ln x)$

Chapter 10

1. Since the population y satisfies $y' = .05y$,
   $y = y_0 e^{.05t} = 1.5e^{.05t}$ billion.
   $y(100) = 1.5e^5$ billion ≈ 223 billion.

3. a. When $y = 1000e^{.1t} = 1,000,000$, $e^{.1t} = 1000$,
   $.1t = \ln 1000, t = 10 \ln 1000 \approx 69$ years.

   b. When $y = 1000(1.1)^t = 1,000,000$,
   $(1.1)^t = 1000, \ln(1.1)^t = t \ln 1.1 = \ln 1000$,
   $t = \frac{\ln 1000}{\ln 1.1} \approx 72\frac{1}{2}$ years.

5. Since $y = y_0(\frac{1}{2})^{t/4.5}$, the fraction *remaining* after 1 billion years
   is $\frac{y}{y_0} = (\frac{1}{2})^{1/4.5}$. The fraction *decayed* is $1 - (\frac{1}{2})^{1/4.5} \approx 14\%$.

7. $1 - (\frac{1}{2})^{\frac{1000}{5730}} \approx 11\%$,

   $5730 \ln .9 / \ln .5 \approx 871$ years,

   $5730 \ln .8 / \ln .5 \approx 1845$ years,

   $5730 \ln .1 / \ln .5 \approx 19035$ years

9. $2 \ln 10000 = 2 \ln 10^4 = 8 \ln 10 \approx 18.4$ days

11. $12/\log_{10} 8 \approx 13.3$ hours. ($y = e^{kt}$. Since $2 = e^{k/3}, e^k = 2^3 = 8, y = 8^t$.)

13. 0

15. 7

17. 1

19. a. $-\infty$ b. $+\infty$

21. 0

Chapter 11

1. $\dot{f} = -32t + 12, \ddot{f} = -32$

3. $\dot{f} = te^t(2 \sin t + t \sin t + t \cos t)$
   $\ddot{f} = 2e^t \sin t + te^t(4 \sin t + 4 \cos t + 2t \cos t)$

5. $y' = (x^4 + 4x^3)e^x$

$y'' = (x^4 + 8x^3 + 12x^2)e^x$

$y''' = (x^4 + 12x^3 + 36x^2 + 24x)e^x$

$y'''' = (x^4 + 16x^3 + 72x^2 + 96x + 24)e^x$

7. $y' = \frac{1}{12}x^2(6\ln x - 11) + \frac{1}{6}x^2$

$y'' = \frac{1}{6}x(6\ln x - 11) + \frac{5}{6}x$

$y''' = \ln x$

$y'''' = 1/x$

9. $f'(x) > 0$ for $1 < x < 4$,

$f'(x) < 0$ for $0 < x < 1$,

$f''(x) > 0$ for $0 < x < 3/2$ or $3 < x < 4$,

$f''(x) < 0$ for $3/2 < x < 3$

11.

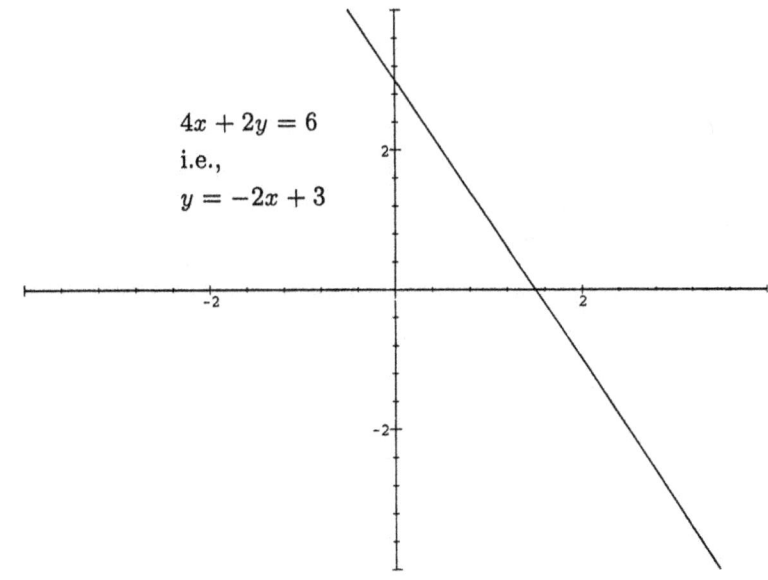

$4x + 2y = 6$

i.e.,

$y = -2x + 3$

(Chapter 11, continued)

13.

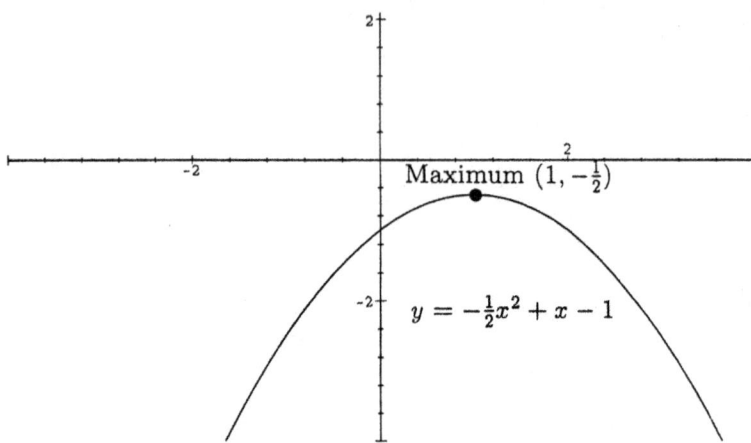

15.

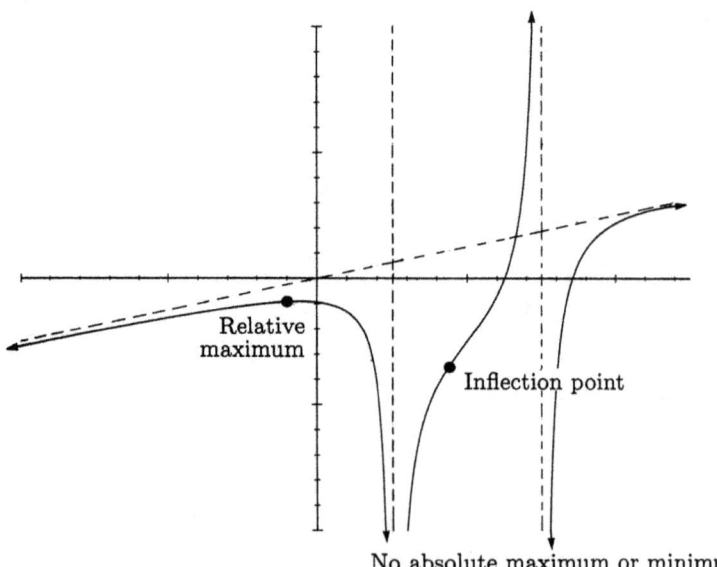

17

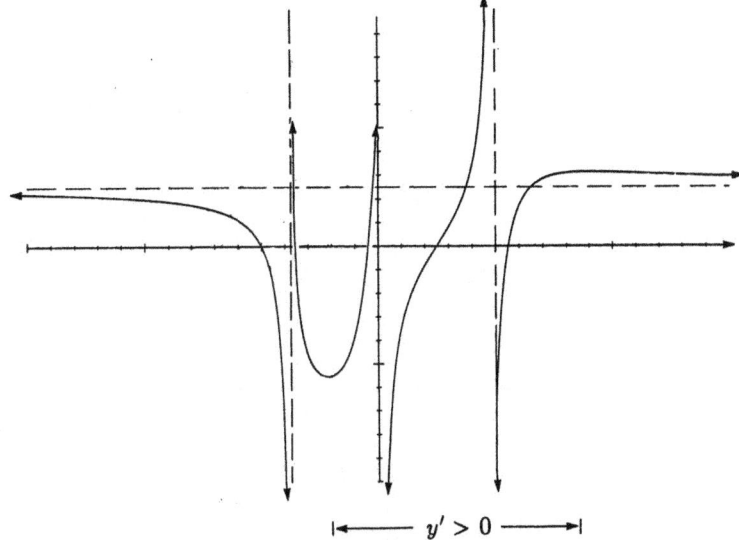

$$\text{|}\!\longleftarrow\!\text{---} \; y' > 0 \;\text{---}\!\longrightarrow\!\text{|}$$

(excluding the endpoints, where $y' = 0$,
and the vertical asymptotes, where y' is undefined)

19.

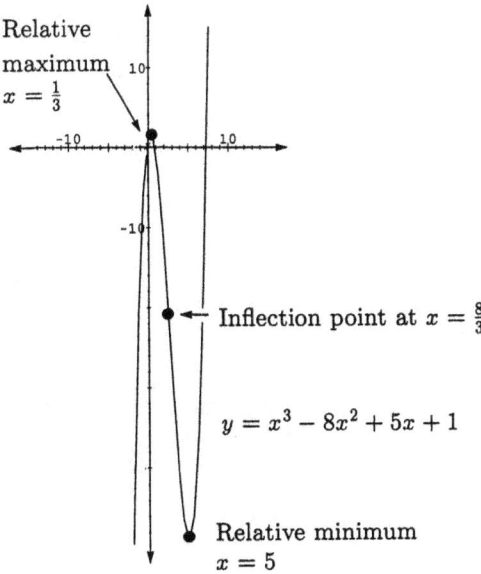

Relative
maximum
$x = \frac{1}{3}$

Inflection point at $x = \frac{8}{3}$

$y = x^3 - 8x^2 + 5x + 1$

Relative minimum
$x = 5$

(Chapter 11, continued)

21.

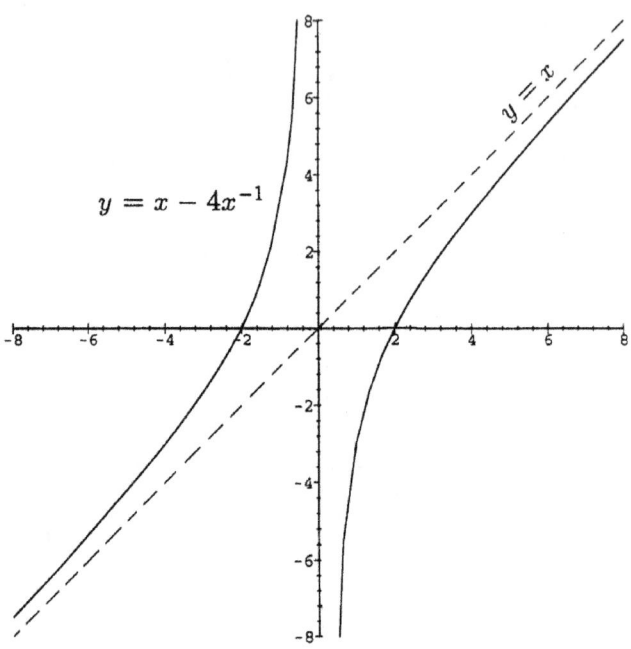

$$y = x - 4x^{-1}$$

No maxima or minima

23.

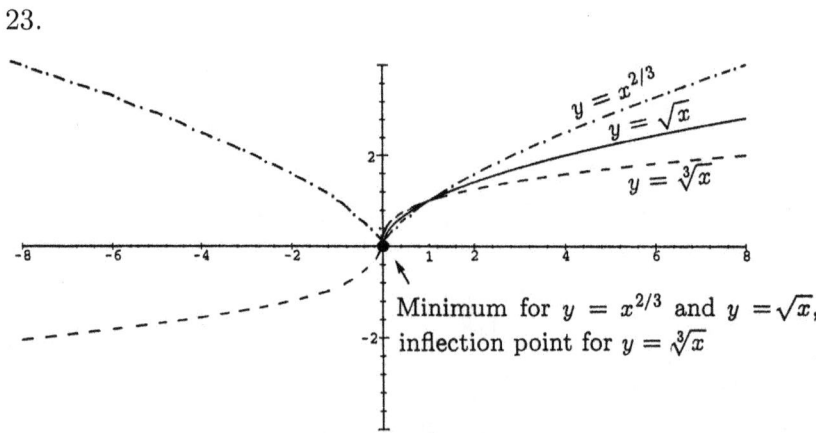

Minimum for $y = x^{2/3}$ and $y = \sqrt{x}$, inflection point for $y = \sqrt[3]{x}$

25.

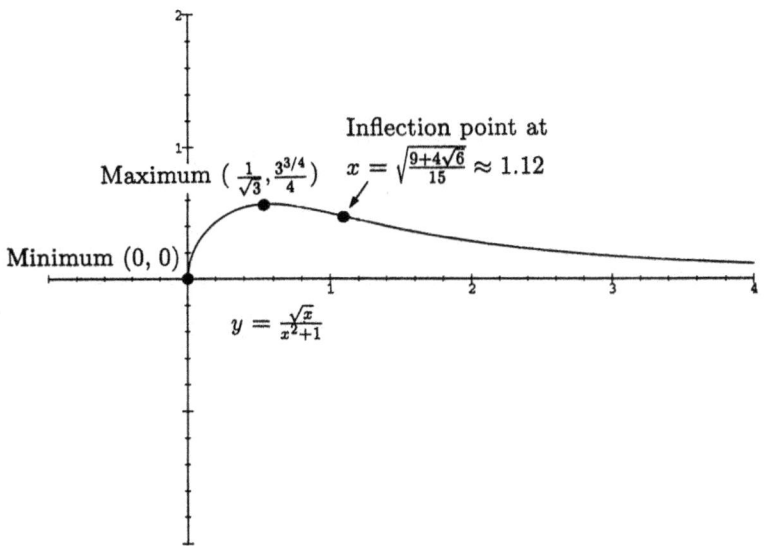

Maximum $\left(\frac{1}{\sqrt{3}}, \frac{3^{3/4}}{4}\right)$

Inflection point at $x = \sqrt{\frac{9+4\sqrt{6}}{15}} \approx 1.12$

Minimum $(0, 0)$

$y = \frac{\sqrt{x}}{x^2+1}$

27.

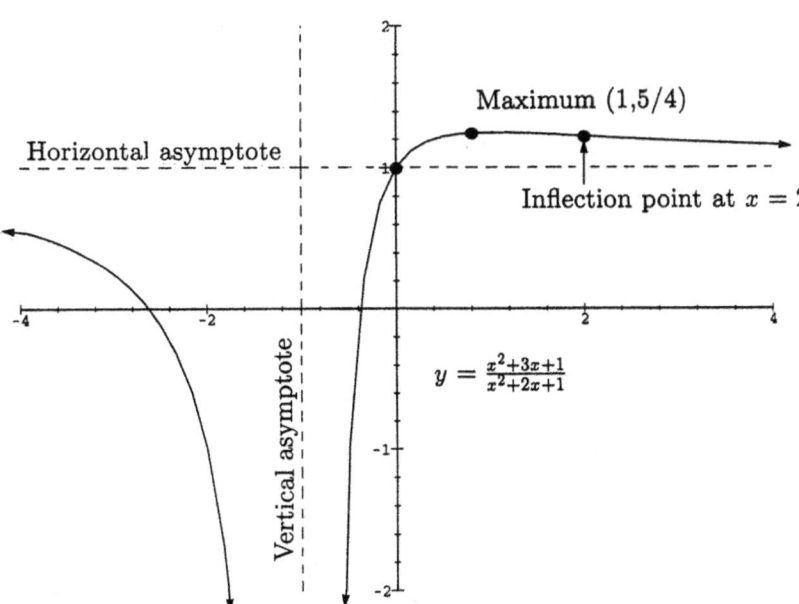

Horizontal asymptote

Maximum $(1, 5/4)$

Inflection point at $x = 2$

Vertical asymptote

$y = \frac{x^2+3x+1}{x^2+2x+1}$

Chapter 12

1. $\frac{1}{6}x^6 +$ C

3. $\frac{9}{4}x^4 - \frac{1}{2}x^2 + 3x +$ C

5. $\frac{1}{2}x^2 +$ C

7. $x +$ C

9. $\frac{2}{3}x^{3/2} +$ C

11. $-\frac{1}{9}x^{-9} +$ C

13. $5\ln|x| +$ C

15. unknown

17. unknown

19. $t^2 - t + 16$

21. $\sin t + t$

23. $f(t) = -16t^2 + 15$

25. a. $h = -16t^2 + 40t + 200$ (antidifferentiate $a = -32$ twice)

  b. $h' = 0 \Rightarrow t = 5/4 \Rightarrow h = 225$

  c. $h = 0 \Rightarrow t = 5$

Chapter 13

1. differentiable except at multiples of $1/2$; continuous everywhere

3. differentiable except at 0, 1, 2, 3; continuous everywhere

5. differentiable except at odd multiples of $\frac{\pi}{2}$ (where $\cos x = 0$); continuous everywhere

7. differentiable except at 0; continuous eveywhere

9. differentiable except at multiples of $1/2$; continuous everywhere (same function as Exercise 13.1)

11. differentiable and continuous on whole domain $\{x \geq 0\}$ (We have not defined real powers of negative numbers.)

13. differentiable and continuous everywhere

15. differentiable and continuous on whole domain $\{x \neq 1, -5\}$

17. differentiable and continuous everywhere

19. differentiable and continuous on whole domain $\{x \neq 0\}$

21. yes (Sums rule 1.4)

yes (Product rule 3.1)

yes where defined $\{g(x) \neq 0\}$ (Quotient rule 3.2)

yes (Chain rule 4.1)

Chapter 14

1. $5\cos 5x$

3. $(3x^2 - 2x)\cos(x^3 - x^2)$

5. $(10\ln 2)2^{10x}$

7. $\cos^{-2} x$

9. $4/(e^x + e^{-x})^2$

11. $8\sin^3 2x \cos 2x \sin^5 3x + 15\sin^4 2x \sin^4 3x \cos 3x$

13. a. $6x\cos(3x^2 + 2)$

b. $-x^{-1}\sin(\ln 5x)$

c. $(\cos(x+2))\ln 3x + x^{-1}\sin(x+2)$

d. $6(\ln x)x^{\ln x - 1}$

e. 0

f. $\dfrac{-(\sin x)(x^2+3) - 2x\cos x}{(x^2+3)^2}$

g. $\cos 3$

h. $-8(9x^2 - 2)\cos(12x^3 - 8x)\sin(12x^3 - 8x)$
$= (-36x^2 + 8)\sin(24x^3 - 16x)$

(Chapter 14, continued)

    i. $-(\cos x)\sin(\sin x)$

    j. $-6x\cos^2(x^2+1)\sin(x^2+1)e^{\cos^3(x^2+1)}$

    k. 0

    l. $2x\cos x - x^2\sin x$

    m. $2(x-1)e^{2x}/x^3$

    n. $(100\ln 2)2^x$

    o. $(3\ln 4)4^{3x}$

    p. $2/x$

15. $f(x) = \frac{x^2+x+1}{x^2+1}$

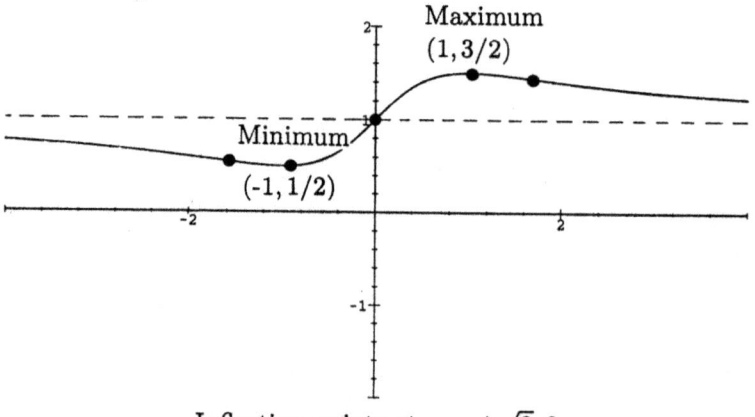

Inflection points at $x = \pm\sqrt{3}, 0$

17. a. $\frac{1}{3}(x^2+2x+1)^{-\frac{2}{3}}(2x+2)$

    b. $\frac{\cos x}{\sin x} = \cot x$

    c. $e^x(\cos x - \sin x)$

    d. $-(\ln t)^{-2}t^{-1}$

    e. $-\sin(t^2)\cdot 2t$

    f. $x^{e^x}(e^x\ln x + e^x/x)$

19. $-\frac{1}{12}t^4 + t^2 + t - 1$

21. It is growing, because $P'(t) = 5e^{t/10} > 0$.

   $P(t) = 500$ when $t = 10 \ln 10 \approx 23$ hrs.

   $P' = \frac{1}{10} P = 7$.

23. $y = 1/e$ is tangent to graph at both maxima.

25. $f(x) = e^{-x}$

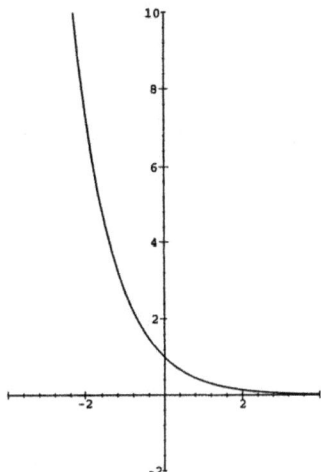

27. a. $\frac{3}{4}x^4 + \frac{2}{3}x^3 + C$

   b. $\frac{2}{3}x^{3/2} + \frac{3}{4}x^{4/3} + C$

   c. $-100\cos\frac{x}{100} + C$

   d. unknown

   e. $100e^{x/100} + C$

   f. unknown

   g. $e^{x^2} + C$

29. On circle $x^2 + y^2 = 1$, by implicit differentiation, $2x + 2yy' = 0 \Rightarrow$
   $y' = -x/y$ at (x, y). Hence at the opposite point $(-x, -y)$, the
   slope $y' = -(-x)/(-y) = -x/y$ is the same.

31. $(x^4)' = \lim_{\Delta x \to 0} \frac{f(x+\Delta x) - f(x)}{\Delta x} = \lim_{\Delta x \to 0} \frac{(x+\Delta x)^4 - x^4}{\Delta x}$

   $= \lim_{\Delta x \to 0} \frac{x^4 + 4x^3\Delta x + 6x^2(\Delta x)^2 + 4x(\Delta x)^3 + (\Delta x)^4 - x^4}{\Delta x}$

   $= \lim_{\Delta x \to 0} (4x^3 + 6x^2\Delta x + 4x(\Delta x)^2 + (\Delta x)^3) = 4x^3$

33. $y' = \frac{5}{2}x^{\frac{3}{2}}$

    $y'(4) = \frac{5}{2} \cdot 8 = 20.$ Line $y = 20x + b$

    When $x = 4$, $y = 4^{\frac{5}{2}} = 32$; $32 = 20 \cdot 4 = b$

    $\Rightarrow b = 32 - 80 = -48$ Line $y = 20x - 48$

35. Extreme cases: $x = 0 \Rightarrow y = 3$

    As $x \to +\infty, y \to +\infty$ **NO MAX**

    Critical cases: $0 = y' = x^2 - x - 2 = (x-2)(x+1) \Rightarrow x = 2$ $(-1 \not\geq 0)$

    $y = \frac{8}{3} - 2 - 4 + 3 = \frac{8}{3} - 3 = \frac{8-9}{3} = -\frac{1}{3}$ **MIN**

37. Extreme cases: $x = 0 \Rightarrow y = 1$ **MAX**

    As $x \to -\infty, y \to -\infty$ **NO MIN**

    Critical cases: $0 = y' = 3x^2 + 2x + 1$, no solution

39. $-\frac{1}{(w-1)^2} - \frac{1}{3w^2}$

41. $\sqrt{\ln x} + 1/2\sqrt{\ln x}$

43. $\frac{2}{3}(-\cos x \sin x + x \cos x^2)(\cos^2 x + \sin x^2)^{-\frac{2}{3}}$

Chapter 15

1. $\int_0^1 x \, dx = \lim_{\Delta x \to 0} \sum x \Delta x = \lim_{n \to \infty} \sum_{k=1}^{n} (\frac{k}{n})(\frac{1}{n})$

    $= \lim_{n \to \infty} \frac{1}{n^2} \sum_{k=1}^{n} k = \lim_{n \to \infty} \frac{1}{n^2} \frac{n(n+1)}{2}$

    $= \lim_{n \to \infty} \frac{1(1+\frac{1}{n})}{2} = \frac{1}{2}$

Chapter 16

1. $\int_0^1 x \, dx = \frac{1}{2}x^2\big]_0^1 = \frac{1}{2}1^2 - \frac{1}{2}0^2 = \frac{1}{2}$

3. 6

5. 10^9

7. $\frac{38}{3}$

9. 396.8

11. $2(\sqrt{3} - \sqrt{2})$

13. $2(e^{\frac{5}{2}} - e^{\frac{3}{2}})$

15. $\sqrt{2}/4$

17. $-\cos x + $ C

19. $-\frac{1}{2}\cos(2x - \frac{\pi}{5}) + $ C

Chapter 17

1. $\frac{34}{3}$

3. .567 (exactly)

5. $\frac{3920}{3}$

7. $2 - \ln 4$

9. $2\sqrt{2} - 2$ $(= \int_0^{\frac{\pi}{4}} (\cos x - \sin x)\, dx + \int_{\frac{\pi}{4}}^{\frac{\pi}{2}} (\sin x - \cos x)\, dx)$

11. $(4\sqrt{2} - 2)/3$

$(= \int_0^{\frac{\pi}{12}} (\cos 3x - \sin 3x)\, dx + \int_{\frac{\pi}{12}}^{\frac{5\pi}{12}} (\sin 3x - \cos 3x)\, dx$

$+ \int_{\frac{5\pi}{12}}^{\frac{\pi}{2}} (\cos 3x - \sin 3x)\, dx)$

13. $\frac{1}{5}e^5 - 5e^{\frac{1}{5}} - \frac{1}{5}e^{25} + 5e$

15. $3/2$

17. -4

Chapter 18

1. $\int (5x + 3)^{99}\, dx = \frac{1}{5}\int (5x + 3)^{99}(5dx) = \frac{1}{5}\frac{(5x+3)^{100}}{100} + $ C

$= \frac{1}{500}(5x + 3)^{100} + $ C

3. $\frac{1}{4}e^{4x-1} + $ C

5. $-\frac{1}{3}\cos(3x + \pi) + $ C

7. $\frac{1}{1000}(5x^2 + 3)^{100} + $ C

9. $\frac{1}{16}\sin(4x^4 - 3) + $ C

11. $\frac{1}{3}(x^2 + 1)^{\frac{3}{2}} + C$

13. $\frac{1}{3}\sin^3 x + C$

15. $-\frac{1}{2} e^{-x^2}\Big]_0^{10} = \frac{1}{2}(1 - e^{-100})$

17. $\frac{1}{3}\ln(x^3 + 1)\Big]_{x=0}^{\sqrt[3]{26}} = \ln 3$

19. $\frac{1}{2}(2x^2 + 1)^{\frac{1}{2}}\Big]_0^2 = 1$

21. $-\frac{1}{2}\int(5 - x^2)^{-\frac{1}{3}}(-2xdx) = -\frac{1}{2}\frac{(5-x^2)^{\frac{2}{3}}}{\frac{2}{3}} = -\frac{3}{4}\sqrt[3]{(5 - x^2)^2} + C$

23. $-\frac{1}{2(x^2+7)} + C$

25. $\frac{1}{12}(x^2 + 3)^6 + C$

Chapter 19

1. $\pi/6$ Since you see $\sqrt{4 - x^2}$, use $x = 2\sin\theta$. When $x = 0$, $\theta = 0$; when $x = 1$, $\sin\theta = 1/2$, $\theta = \pi/6$.

$\sqrt{4 - x^2}$ becomes $\sqrt{4 - (2\sin\theta)^2} = \sqrt{4 - 4\sin^2\theta}$

$= \sqrt{4}\sqrt{1 - \sin^2\theta} = 2\cos\theta$, dx becomes $2\cos\theta\,d\theta$, and

$\int_{x=0}^1 \frac{dx}{\sqrt{4-x^2}} = \int_{\theta=0}^{\pi/6} \frac{2\cos\theta\,d\theta}{2\cos\theta} = \int_{\theta=0}^{\pi/6} 1 \cdot d\theta$

$= \theta]_{\theta=0}^{\pi/6} = \frac{\pi}{6} - 0 = \frac{\pi}{6}$

3. $3\pi/4 + 9\sqrt{3}/8$. Use $x = 3\sin\theta$.

$\int_{x=0}^{3/2} \sqrt{9 - x^2}\,dx = \int_{\theta=0}^{\pi/6}(3\cos\theta)(3\cos\theta d\theta) = 9\int_{\theta=0}^{\pi/6}\cos^2\theta\,d\theta$

$= 9\int_{\theta=0}^{\pi/6}(\frac{1}{2} + \frac{1}{2}\cos 2\theta)\,d\theta = 9\left[\frac{\theta}{2} + \frac{1}{4}\sin 2\theta\right]_{\theta=0}^{\pi/6}$

$= 9[(\frac{\pi}{12} + \frac{1}{4}\sin\frac{2\pi}{6}) - (0 + 0)] = \frac{3\pi}{4} + \frac{9}{4}\frac{\sqrt{3}}{2} = \frac{3\pi}{4} + \frac{9\sqrt{3}}{8}$

5. $\sqrt{3}/2$

First way: $\int_{x=0}^{3/2} \frac{x\,dx}{\sqrt{3-x^2}} = -\frac{1}{2}\int_{x=0}^{3/2}(3-x^2)^{-\frac{1}{2}}(-2x)\,dx$

$= -\frac{1}{2}\left[\frac{(3-x^2)^{\frac{1}{2}}}{\frac{1}{2}}\right]_{x=0}^{3/2} = -\sqrt{3-x^2}\Big]_{x=0}^{3/2}$

$= -\sqrt{3-(\frac{3}{2})^2} - (-\sqrt{3-0^2}) = -\sqrt{\frac{3}{4}} + \sqrt{3} = \sqrt{3}[-\frac{1}{2}+1] = \frac{\sqrt{3}}{2}$

Second way: Let $x = \sqrt{3}\sin\theta$, $dx = \sqrt{3}\cos\theta\,d\theta$

$\int_{x=0}^{3/2}\frac{x\,dx}{\sqrt{3-x^2}} = \int_{\theta=0}^{\pi/3}\frac{(\sqrt{3}\sin\theta)(\sqrt{3}\cos\theta\,d\theta)}{\sqrt{3}\cos\theta}$

$= -\sqrt{3}\cos\theta\Big]_{\theta=0}^{\pi/3} = -\sqrt{3}(\frac{1}{2}) - (-\sqrt{3}\cdot1) = \frac{\sqrt{3}}{2}$

7. $\sqrt{3}$

9. Let $u = 7x + 3$. Integral $= \int \frac{\frac{u-3}{7}\frac{du}{7}}{u} = \frac{1}{49}\int(1 - \frac{3}{u})\,du$

$= \frac{1}{49}[u - 3\ln|u|] + C = \frac{1}{49}[7x + 3 - 3\ln|7x + 3|] + C$

$= \frac{x}{7} - \frac{3}{49}\ln|7x + 3| + C'$ $(C' = C + \frac{3}{49})$

11. $\frac{\pi}{3\sqrt{3}} + \frac{1}{4}$ (Use $x = \sqrt{\frac{2}{3}}\sin\theta$.)

Chapter 21

1. $3\sec^2 x$

3. $3x^2\sec^2 x^3$

5. $\frac{1}{2\sqrt{x}}\sec^2\sqrt{x}$

7. $(2x + 3)\sec(x^2 + 3x - 7)\tan(x^2 + 3x - 7)$

9. $\cot x - x\csc^2 x$

11. $\cot x$

13. $\tan^4 x$

15. $16/65$

17. $-\pi/3$

19. $5\pi/6$

21. $\pi/3$

23. $\frac{2}{5}\sqrt{6}$

25. $\sqrt{7}/3$

27. $5/\sqrt{1-x^2}$

29. $5(\sin^{-1}x)^4/\sqrt{1-x^2}$

31. $3/(1+9x^2)$

33. $1/2\sqrt{x}(1+x)$

35. $e^{\tan^{-1}x}/(1+x^2)$

37. Implicit differentiation of $\tan y = x$ yields $(\sec^2 y)y' = 1$,

   $y' = 1/(1+\tan^2 y) = 1/(1+x^2)$

39. $\pi/6$

41. $\pi/2$

43. $\frac{1}{2}\tan^{-1}(2x-3)+$ C

45. $\tan^{-1}x^2+$ C

Chapter 22

1. I guess $V \approx$ volume ball radius $3 = \frac{4}{3}\pi 3^3 \approx 100$.

   $V = \int_{x=-2}^{2}\pi(4-x^2)^2 dx = 512\pi/15 \approx 107$.

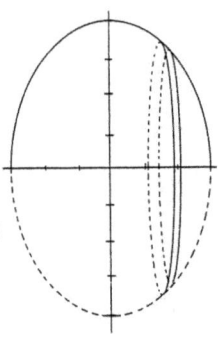

3. Guess: $\pi(1^2)4 = 12+$. $V = 16\pi/3 \approx 16.8$.

5. Guess 8 or 9.

   $L = \int_{x=1}^{4}\sqrt{1+\frac{9}{4}x}\,dx = \frac{8}{27}\left[10^{3/2}-(13/4)^{3/2}\right] \approx 7.6$

7. $\frac{1}{2\pi}\int_{x=0}^{2\pi}\cos^2 x dx = \frac{1}{2\pi}\int_{x=0}^{2\pi}(\frac{1}{2}+\frac{1}{2}\cos 2x)\,dx$

   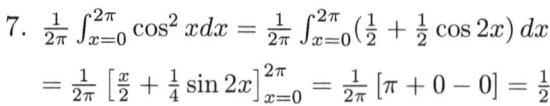

   $= \frac{1}{2\pi}\left[\frac{x}{2}+\frac{1}{4}\sin 2x\right]_{x=0}^{2\pi} = \frac{1}{2\pi}\left[\pi+0-0\right] = \frac{1}{2}$

9. $(e^{100}-1)/100$.   Less

Chapter 23

Warning: Different correct methods can yield correct answers hard to recognize as equivalent to the given solutions.

1. Use Table 5 with $u = 3\theta - \frac{\pi}{12}$

$\int \cos^2(3\theta - \frac{\pi}{12}) \, d\theta = \frac{1}{3} \int \cos^2(3\theta - \frac{\pi}{12})(3d\theta)$

$= \frac{1}{3} \int \cos^2 u \, du = \frac{1}{3}[\frac{1}{2}u + \frac{1}{4} \sin 2u + C]$

$= \frac{1}{6}u + \frac{1}{12} \sin 2u + C$

$= \frac{1}{6}(3\theta - \frac{\pi}{12}) + \frac{1}{12} \sin(6\theta - \frac{\pi}{6}) + C$

$= \frac{\theta}{2} + \frac{1}{12} \sin(6\theta - \frac{\pi}{6}) + C$ (The $-\frac{\pi}{12}$ just changes the constant C.)

3. By Table 17(2) with $n = 0$, $m = 6$,

$\int \cos^6 \theta \, d\theta = \frac{\sin \theta \cos^5 \theta}{6} + \frac{5}{6} \int \cos^4 \theta \, d\theta$

which by Table 10 becomes

$\frac{\sin \theta \cos^5 \theta}{6} + \frac{5}{16}\theta + \frac{5}{24} \sin 2\theta + \frac{5}{192} \sin 4\theta + C$

5. $-\frac{1}{8} \sin^7 \theta \cos \theta - \frac{7}{48} \sin^5 \theta \cos \theta + \frac{35}{128}\theta - \frac{35}{192} \sin 2\theta + \frac{35}{1536} \sin 4\theta + C$

7. Use Table 17 with $n = m = 2$ to get

$-\frac{\sin \theta \cos^3 \theta}{4} + \frac{1}{4} \int \cos^2 \theta d\theta + C.$

Now use Table 5 to get

$-\frac{1}{4} \sin \theta \cos^3 \theta + \frac{1}{8}\theta + \frac{1}{16} \sin 2\theta + C.$

Another method is to start by simplifying to

$\frac{1}{4} \int \sin^2 2\theta \, d\theta = \frac{1}{8} \int \sin^2 2\theta (2d\theta)$

and using Table 6 to get

$\frac{1}{8}[\frac{1}{2}(2\theta) - \frac{1}{4} \sin 4\theta] = \frac{1}{8}\theta - \frac{1}{32} \sin 4\theta + C.$

It takes some effort to see that these two answers agree:

$-\frac{1}{4} \sin \theta \cos^3 \theta + \frac{1}{8}\theta + \frac{1}{16} \sin 2\theta = -\frac{1}{8}(2 \sin \theta \cos \theta) \cos^2 \theta + \frac{1}{8}\theta + \frac{1}{16} \sin 2\theta$

$= \frac{1}{8}\theta - \frac{1}{8} \sin 2\theta(\frac{1}{2} + \frac{1}{2} \cos \theta) + \frac{1}{16} \sin 2\theta$

$= \frac{1}{8}\theta - \frac{1}{16} \sin 2\theta \cos 2\theta = \frac{1}{8}\theta - \frac{1}{32} \sin 4\theta$

9. $-\frac{1}{3}x \cos 3x + \frac{1}{9} \sin 3x + C$

(Use Table 18 with $n = 1, a = 3$)

11. $\frac{1}{2}\tan^2 x + \ln|\cos x| + C$ (Table 20)

13. $\frac{1}{2}\left(\frac{1}{\sqrt{21}}\tan^{-1}\frac{x}{\sqrt{21}}\right) + C$

15. $\frac{x}{4(2+x^2)} + \frac{1}{4\sqrt{2}}\tan^{-1}\frac{x}{\sqrt{2}} + C$ (Table 22)

17. $\int \frac{dx}{((x+1)^2+1)^2} = \frac{x+1}{2((x+1)^2+1)} + \frac{1}{2}\tan^{-1}(x+1) + C$

   $= \frac{x+1}{2(x^2+2x+2)} + \frac{1}{2}\tan^{-1}(x+1) + C$

19. $\ln(x + \sqrt{x^2+7}) + C$

21. $(x^5 - 5x^4 + 20x^3 - 60x^2 + 120x - 120)e^x + C$

23. $\frac{x-\frac{1}{2}}{22(\frac{11}{4}-(x-\frac{1}{2})^2)} + \frac{1}{22\sqrt{11}}\ln\left|\frac{x-\frac{1}{2}+\frac{\sqrt{11}}{2}}{x-\frac{1}{2}-\frac{\sqrt{11}}{2}}\right| + C$

Chapter 24

1. First $\frac{1}{x^2-1} = \frac{1}{(x+1)(x-1)} = \frac{-\frac{1}{2}}{x+1} + \frac{\frac{1}{2}}{x-1}$.

   Hence $\int \frac{dx}{x^2-1} = -\frac{1}{2}\ln|x+1| + \frac{1}{2}\ln|x-1| + C$

   $= \frac{1}{2}\ln\left|\frac{x-1}{x+1}\right| + C$

3. $\frac{4}{5}\ln|x-4| + \frac{1}{5}\ln|x+1| + C$

5. $-\frac{1}{10}\ln|x+1| - \frac{3}{2}\ln|x-3| + \frac{13}{5}\ln|x-4| + C$

   (Denominator factors as $(x+1)(x-3)(x-4)$.)

7. $\frac{1}{\sqrt{21}}\ln\left|\frac{x+\frac{5}{2}-\frac{\sqrt{21}}{2}}{x+\frac{5}{2}+\frac{\sqrt{21}}{2}}\right| + C$

   By the quadratic formula, the roots of the denominator are $\frac{-5\pm\sqrt{21}}{2}$. Hence the denominator factors as $(x+\frac{5+\sqrt{21}}{2})(x+\frac{5-\sqrt{21}}{2})$.

9. Let $u = x$, $dv = e^x dx$; $du = dx$, $v = e^x$.

   $\int u\,dv = uv - \int v\,du = xe^x - \int e^x dx$

   $= xe^x - e^x + C = (x-1)e^x + C$

11. $\frac{1}{2}x\sin 2x + \frac{1}{4}\cos 2x + C$

13. First use of integration by parts yields

$-x^2 \cos x - \int(-\cos x)(2x dx) = -x^2 \cos x + \int 2x \cos x dx$

Second use yields the answer:

$-x^2 \cos x + 2x \sin x + 2 \cos x + C$

15. First use of integration by parts yields:

$I = \int e^x \cos x dx = e^x \sin x - \int e^x \sin x dx$

Second use yields:

$I = e^x \sin x + e^x \cos x - \int e^x \cos x dx = e^x \sin x + e^x \cos x - I + C$

$2I = e^x \sin x + e^x \cos x + C$

$I = \frac{1}{2} e^x (\sin x + \cos x) + C$

17. Let $u = x$, $du = dx$, $dv = e^x \cos x dx$.

By Exercise 15, $v = \frac{1}{2} e^x (\sin x + \cos x)$

Answer $= \frac{1}{2} x e^x (\sin x + \cos x) - \frac{1}{2} \int e^x (\sin x + \cos x)$

$= \frac{1}{2} x e^x (\sin x + \cos x) - \frac{1}{2} e^x \sin x + C$ by 15 and 16.

19. $x \sin^{-1} x + \sqrt{1 - x^2} + C$

21. By parts, $\int_0^1 x f''(x) dx = x f'(x)]_{x=0}^1 - \int_0^1 f'(x) dx$

$= 0 - (f(1) - f(0)) = f(0) - f(1)$.

Chapter 25

1. $\frac{1}{12}[1 + 4 \cdot \frac{4}{5} + 2 \cdot \frac{2}{3} + 4 \cdot \frac{4}{7} + \frac{1}{2}] \approx .6933$

Exact $= \ln 2 \approx .6931$

3. 8/3 (exact)

5. $6\frac{5}{12} \approx 6.42$

Exact $= 6\frac{2}{5} = 6.40$.

7. $\approx \frac{1}{6}[1 + 4(.7788) + 2(.3679) + 4(.1054) + .0183] \approx .8818$

Antiderivative unknown

9. n 2   4   6   10   12   14   16   20

   $\approx$ 2.741 2.408 2.342 2.312 2.308 2.306 2.3046 2.3035

   Exact: ln 10 $\approx$ 2.303.   See 25.11

11. 54 (n must be even)

Chapter 26

1. $\frac{\sqrt{3}-1}{2}$

3. $(x-1)e^x + C$

5. 1

7. $1/4 \ln 2$

9. $\frac{1}{2}\ln(x^2+4) + C$

11. $-(4-x^2)^{\frac{1}{2}} + C$

13. a. $\frac{1}{4}x^4 + x^3 + x^2 + C$

    b. $\frac{1}{2}(e^4-1)$

    c. $-x\cos x + \sin x + C$

    d. $\pi/4\sqrt{5}$

    e. $\frac{1}{2}\ln(x^2+5) + C$

    f. $\frac{1}{2}\tan^2 x + C$

    g. $x\arcsin x + \sqrt{1-x^2} + C$

    h. $\frac{\pi}{2} - \frac{1}{\sqrt{2}}(\frac{\pi}{4}+1)$

15. $\frac{1}{2}\ln^2 x + C$

17. $\frac{1}{3}\tan^3 x + C$

19. $\frac{3}{2}\ln(25+x^2) + C$

21. $x\arctan x - \frac{1}{2}\ln(1+x^2) + C$

23. $4\sin^{-1}\frac{x}{3} + C$

25. $-2x^{-2} + \ln|x| + C$

27. $\frac{1}{2}e^{x^2} + C$

29. $1/8$, $3/5$, $1/8$

31. a. $\int_{x=0}^{4}((6x - x^2) - (x^2 - 2x))dx = 64/3$

The limits of integration $x = 0, 4$ come from where the two parabolas meet:

$6x - x^2 = x^2 - 2x$

$0 = 2x^2 - 8x = 2x(x - 4) \quad x = 0, 4$

b. $125/2$

33. a. We cannot do this one.

b. Recognize $\int \sin u \, du$.

c. Use $\sin^2 \theta = \frac{1}{2} - \frac{1}{2} \cos 2\theta$.

d. Recognize $\int u^2 \, du$.

e. Hard (start by expressing $\sin 3x$ in terms of $\sin x$ and $\cos x$).

f. We cannot do this one.

g. Recognize $\int u^2 \, du$.

h. Use integration by parts.

35. a. x

b. $.83$

c. undefined

d. $\frac{5\pi}{8}$

e. $\frac{3\pi}{8}$

f. $-2\pi \csc 2\pi x \cot 2\pi x$

g. $-\frac{1}{\sqrt{1-x^2}}$

37. a. $\sec^2 x$, $-\ln|\cos x| + C$

b. $\sec x \tan x$, $\ln|\sec x + \tan x| + C$

c. $2\sec^2 x \tan x$, $\tan x + C$

d. $2\tan x \sec^2 x$, $\tan x - x + C$

e. $-\frac{1}{2}(a^2 - x^2)^{-\frac{3}{2}}(-2x)$, $\sin^{-1}\frac{x}{a} + C$

f. $\frac{1}{\sqrt{1-x^2}}$, $x\sin^{-1}x + \sqrt{1-x^2} + C$

g. $(\ln 2)2^x$, $\frac{1}{\ln 2}2^x + C$

h. $\frac{1}{x}$, $x(\ln ax - 1) + C$

39. Area or volume of m-dimensional balls.

41. a. Continuous growth rate k.

b. Exponential decay with halflife h.

c. Arclength when x, y given as functions of some other parameter t.

d. Volume sliced into washers.

e. Integrating product.

Chapter 27

1. Converges to $\frac{1}{4}$ (geometric series)

3. Diverges to ∞ by p-test.

5. Converges by p-test.

7. Diverges by oscillation.

9. Diverges by ratio test to ∞

11. Converges by ratio test.

13. Converges by comparison with $\sum \frac{1}{n^2}$

15. Converges by comparison with $\sum \frac{1}{n^2}$.

Chapter 28

1. $1 + 2x + \frac{4x^2}{2!} + \frac{8x^3}{3!} + \dots$

3. $1 + x^2 + x^4 + x^6 + \dots$

5. $x - \frac{x^2}{2} + \frac{x^3}{3} - \frac{x^4}{4} + \dots$

7. $x + \frac{1}{3}x^3$

9. 1.225

Chapter 29

1. Separable:

$ydy = \cos x dx, \quad \int y\,dy = \int \cos x\,dx$

$\frac{1}{2}y^2 = \sin x + \text{C}, \quad y^2 = 2\sin x + \text{C} \qquad \text{(new C)}$

$y = \pm\sqrt{2\sin x + \text{C}} \quad [\text{ Not } \pm\sqrt{2\sin x} + \text{C}]$

3. $y = \tan(x + \frac{1}{3}x^3 + \text{C})$

$\left[\int \frac{dy}{1+y^2} = \tan^{-1} y = x + \frac{1}{3}x^3 + \text{ C} \right]$

5. 1st order linear $\rho = e^{2\ln|x|} = x^2$.

$y = x^{-2}(\int x^5\,dx + \text{ C}) = x^{-2}(\frac{1}{6}x^6 + \text{C})$

$y = \frac{1}{6}x^4 + \text{ C}x^{-2}$

7. $y = \pm\sqrt{x^2 + \text{ C}}$.

9. 1st order linear with $p(x) = -\frac{1}{x}$, $q(x) = x$

$\rho(x) = e^{-\ln|x|} = |x|^{-1}$ or just use $\rho(x) = x^{-1}$.

$y = \rho^{-1}(\int \rho q dx + \text{ C}) = x(\int 1\,dx + \text{ C}) = x(x + \text{ C})$

$y = x^2 + \text{ C}x$

11. $\frac{1}{2}y^2 + y = \frac{1}{3}x^3 - x + \text{ C}$

or by the quadratic formula $y = -1 \pm \sqrt{\frac{2}{3}x^3 - 2x + \text{ C}}$ (new C)

13. $y = \frac{x(\ln|x| + \text{ C})}{x+2}$ (1st order linear)

15. $y = -\frac{2}{3}x + \frac{10}{9} + \text{ C}e^{-\frac{3}{5}x}$

Chapter 29 continued

17. First solve to get $y = \pm\sqrt[4]{\frac{4}{3}x^3 + \text{ C}}$.

    To get C, use that when $x = 1$, $y = 1$:

    $1 = \pm\sqrt[4]{\frac{4}{3} + \text{ C}}$ so we must have + sign and $\text{C} = -\frac{1}{3}$;

    $y = \sqrt[4]{\frac{4}{3}x^3 - \frac{1}{3}}$

19. $y = -\frac{4}{7}x^2 + \frac{16}{49}x - \frac{32}{343}(1 - e^{-\frac{7}{2}x})$

21. Separable $\frac{dA}{A} = \frac{1}{2}z^{\frac{1}{2}}dz$

    $\ln|A| = \frac{1}{3}z^{\frac{3}{2}} + \text{C}_1$

    $|A| = e^{\frac{1}{3}z^{\frac{3}{2}} + \text{ C}_1} = e^{\frac{1}{3}z^{\frac{3}{2}}}e^{\text{C}_1}$

    $A = \pm e^{\text{C}_1}e^{\frac{1}{3}z^{\frac{3}{2}}} = \text{C}_2 e^{\frac{1}{3}z^{\frac{3}{2}}}$

23. First order linear: $f' - \frac{1}{z}f = -.01z^2$

    $\rho = e^{-\ln|z|} = \frac{1}{|z|} = \pm\frac{1}{z}$ Use $\rho = \frac{1}{z}$

    $f = \rho^{-1}\int \rho q = z \int \frac{1}{z}(-.01z^2)\,dz$

    $= -.01z(\frac{1}{2}z^2 + \text{ C}_1) = -\frac{1}{200}z^3 + \text{C}_2 z$

25. The differential equation is $\frac{dV}{dt} = -kA$.

    $\frac{d}{dt}(\frac{4}{3}\pi r^3) = -k(4\pi r^2)$

    $4\pi r^2\frac{dr}{dt} = -4\pi kr^2$

    $\frac{dr}{dt} = -k$; $r = -kt + \text{ C}$.

    Plugging in given values yields $1 = \text{C}$, $.5 = -k + \text{ C}$,

    so that $\text{C} = 1$, $k = .5$, $r = -.5t + 1$.

    $r = 0$ and the moths get my sweater when $t = 2$, i.e. July 1 (or maybe July 2).

27. a. $\frac{dp}{p(a-bp)} = dt$, $\left(\frac{\frac{1}{a}}{p} + \frac{\frac{b}{a}}{a-bp}\right)dp = dt$

    $\frac{1}{a}\ln|p| - \frac{1}{a}\ln|a - bp| = t + \text{ C}_1$

    $\ln\left|\frac{p}{a-bp}\right| = at + \text{C}_2$

    $\frac{p}{a-bp} = \pm e^{at + \text{C}_2} = \text{C}_3 e^{at}$

$p = a\ C_3 e^{at} - bp\ C_3 e^{at}$

$p(1 + b\ C_3 e^{at}) = a\ C_3 e^{at}$

$p = \frac{a\ C_3 e^{at}}{1 + b\ C_3 e^{at}} = \frac{a}{C_3^{-1} e^{-at} + b} = \frac{a}{b + ce^{-at}}$

b. $p_\infty = a/b$

c. $c = 7.877 \times 10^{-9}$; 39.1, 122.0, 180, 197 millions

29. $|x| > |y|$

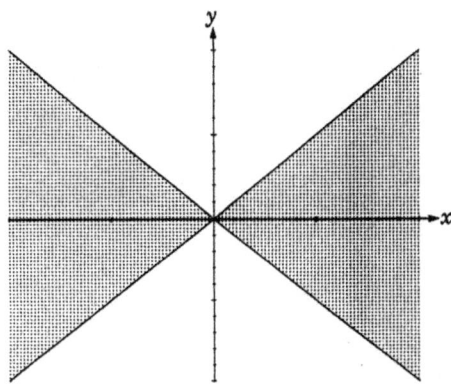

31. a. Amount y. Constant volume $V = V_0$.

$y' = kr - \frac{y}{V_0} \cdot \frac{r}{2}$

Separable. Get $y = 2kV_0 + Ce^{-\frac{r}{2V_0}t}$

$c = 2k + Ce^{-\frac{r}{2V_0}t}$ (new C).

Since $c(0) = k$, $C = -k$.

Answer: $c(t) = k(2 - e^{-\frac{r}{2V_0}t}) \to 2k$ as $t \to \infty$

b. Here $y' = kr - \frac{y}{V}\frac{r}{2} = kr - \frac{y}{V_0 + \frac{r}{2}t} \cdot \frac{r}{2}$

1st order linear: $y' + \frac{1}{t + \frac{2V_0}{r}}y = kr$

$\rho = t + \frac{2V_0}{r}$

$y = (t + \frac{2V_0}{r})^{-1} \left(\int(krt + 2kV_0)dt + C \right)$

$y = (t + \frac{2V_0}{r})^{-1} \left(\frac{r}{2}kt^2 + 2kV_0t + C \right)$

$c = \frac{y}{V} = \frac{\frac{r}{2}kt^2 + 2kV_0t + C}{(t + \frac{2V_0}{r})(V_0 + \frac{r}{2}t)}$

Since $c(0) = 0$, $C = 0$

$$c = k \frac{\frac{r}{2}t^2 + 2V_0 t}{\frac{r}{2}t^2 + 2V_0 t + \frac{2V_0^2}{r}} \to k \text{ as } t \to \infty$$

c. $c(t) = k$ (of course)

Chapter 30

1. The characteristic equation $r^2 - 5r + 6 = 0$ factors as $(r - 2)(r - 3) = 0$, with roots $r = 2, 3$. Hence
$$y = C_1 e^{2x} + C_2 e^{3x}$$

3. $y = (C_1 + C_2 x)e^{3x}$

5. $y = C_1 \cos 3x + C_2 \sin 3x$

7. $y = e^{-\frac{1}{3}x}(C_1 \cos \frac{\sqrt{2}}{3}x + C_2 \sin \frac{\sqrt{2}}{3}x)$

9. $y = 0$ (obviously works)

11. $y = \frac{13}{2}e^{3x} - \frac{9}{2}e^{5x}$

13. $y \to 0$ (increasingly damped, like mass on spring with $m = 5$, $k = 3$, and damping $c = x \to +\infty$)

Chapter 31

1. $f_x = 2xy + y^3$, $f_y = x^2 + 3xy^2$

3. $f_x = 3(x + y)^2$, $f_y = 3(x + y)^2$

5. $f_x = -yx^{-2}e^{y/x}$, $f_y = x^{-1}e^{y/x}$

7. $f_x = \frac{4xy^2}{(x^2+y^2)^2}$, $f_y = -\frac{4x^2 y}{(x^2+y^2)^2}$

9. $f_x = 3\cos 2y$, $f_y = -6x \sin 2y$

11. $f_x = 2xy^2 z^2 + 4x$, $f_y = 2x^2 yz^2 + 9y^2$, $f_z = 2x^2 y^2 z + 16z^3$

13. $\partial V/\partial T = nR/p$, $\partial V/\partial p = -nRT/p^2$

15. $f_a = 2a - 2b\cos\theta$, $f_b = 2b - 2a\cos\theta$, $f_\theta = 2ab\sin\theta$

17. $z_{xx} = 12x^2 y^2$, $z_{xy} = 8x^3 y$, $z_{yy} = 2x^4$

19. $z_{xx} = 20(19x^2 + y^2)(x^2 + y^2)^8$

$z_{xy} = 360xy(x^2 + y^2)^8$

$z_{yy} = 20(x^2 + 19y^2)(x^2 + y^2)^8$

21. Note $z = 3\ln x + 5\ln y$

$z_{xx} = -3/x^2,\ z_{xy} = 0,\ z_{yy} = -5/y^2$

Chapter 32

1. INSIDE $= \frac{1}{2}x^2y^2]_{y=2}^x = \frac{1}{2}x^4 - 2x^2$

OUTSIDE $= \frac{1}{10}x^5 - \frac{2}{3}x^3]_1^2 = (\frac{32}{10} - \frac{16}{3}) - (\frac{1}{10} - \frac{2}{3})$

$= \frac{96-160-3+20}{30} = -\frac{47}{30}$

3. 48

5. $23/6$

7. $\pi^3/24$

9. $\pi/12$

11. $\pi/18$

13. $1/24$

Chapter 33

1. $(1, 2), (1, -2)$

3. $(0, 0), (-18, 6)$

$0 = f_x = 2x + 6y \Rightarrow x = -3y$

$0 = f_y = 6x + 3y^2 = -18y + 3y^2 = 3y(-6 + y) \Rightarrow y = 0$ or 6.

5. $(0, 0), (1, 1), (-1, -1)$

$0 = f_x = 4x^3 - 4y \Rightarrow y = x^3$

$0 = f_y = 4y^3 - 4x \Rightarrow x = y^3 = x^9$

$0 = x^9 - x = x(x^8 - 1) \Rightarrow x = 0, 1,$ or $-1.$ $y = x^3 = 0, 1,$ or $-1.$

7. The whole line $\{y = x\}$.

Chapter 34

1. No absolute maxima (since $f(x, y) \to +\infty$ as $x, y \to \pm\infty$), absolute minimum of 0 at (0,0).

3. No max, min of 1 at (1,0), which comes from checking the boundary $x = 1$, where $f = 1 + y^2$.

5. Max of 13 at (2,3), min of 0 at (0,0)

7. Captain Jones likes (0,3), where it is 91°; Artudeetu likes (0,0), where it is 100°.

Greek Letters

Greek Letters Used in This Text

Lowercase	Capital	Name
α		alpha
δ	Δ	delta
ϵ		epsilon
θ		theta
π		pi
ρ		rho
	Σ	sigma

Greek Alphabet

α	A	alpha	ν	N	nu
β	B	beta	ξ	Ξ	xi
γ	Γ	gamma	o	O	omicron
δ	Δ	delta	π	Π	pi
ϵ	E	epsilon	ρ	P	rho
ζ	Z	zeta	σ	Σ	sigma
η	H	eta	τ	T	tau
θ	Θ	theta	υ	Υ	upsilon
ι	I	iota	ϕ	Φ	phi
κ	K	kappa	χ	X	chi
λ	Λ	lambda	ψ	Ψ	psi
μ	M	mu	ω	Ω	omega

Index

INTEGRAL TABLES

Trigonometric Formulas

1. $\displaystyle\int \cos u\,du = \sin u$

2. $\displaystyle\int \sin u\,du = -\cos u$

3. $\displaystyle\int \tan u\,du = -\ln|\cos u|$

4. $\displaystyle\int \sec u\,du = \ln|\sec u + \tan u|$

5. $\displaystyle\int \cos^2 u\,du = \frac{1}{2}u + \frac{1}{4}\sin 2u$

6. $\displaystyle\int \sin^2 u\,du = \frac{1}{2}u - \frac{1}{4}\sin 2u$

7. $\displaystyle\int \tan^2 u\,du = \tan u - u$

8. $\displaystyle\int \cos^3 u\,du = \sin u - \frac{1}{3}\sin^3 u$

9. $\displaystyle\int \sin^3 u\,du = -\cos u + \frac{1}{3}\cos^3 u$

10. $\displaystyle\int \cos^4 u\,du = \frac{3}{8}u + \frac{1}{4}\sin 2u + \frac{1}{32}\sin 4u$

11. $\displaystyle\int \sin^4 u\,du = \frac{3}{8}u - \frac{1}{4}\sin 2u + \frac{1}{32}\sin 4u$

12. $\displaystyle\int \cos u \sin u\,du = -\frac{1}{4}\cos 2u$

13. $\displaystyle\int_0^{\pi/2} \cos^n u\, du = \int_0^{\pi/2} \sin^n u\, du$

$\displaystyle = \frac{1 \cdot 3 \cdot 5 \ldots (n-1)}{2 \cdot 4 \cdot 6 \ldots n}\frac{\pi}{2}$ if $n \geq 2$ is even

$\displaystyle = \frac{2 \cdot 4 \cdot 6 \ldots (n-1)}{1 \cdot 3 \cdot 5 \ldots n}$ if $n \geq 3$ is odd

14. $\displaystyle\int \sin^{-1} ax\, dx = x \sin^{-1} ax + \frac{1}{a}\sqrt{1 - a^2 x^2}$

15. $\displaystyle\int \cos^{-1} ax\, dx = x \cos^{-1} ax - \frac{1}{a}\sqrt{1 - a^2 x^2}$

16. $\displaystyle\int \tan^{-1} ax\, dx = x \tan^{-1} ax - \frac{1}{2a}\ln(1 + a^2 x^2)$

Reduction Formulas

17. $\displaystyle\int \sin^n u \cos^m u\, du = -\frac{\sin^{n-1} u \cos^{m+1} u}{m+n} + \frac{n-1}{m+n}\int \sin^{n-2} u \cos^m u\, du$

or $\displaystyle\int \sin^n u \cos^m u\, du = \frac{\sin^{n+1} u \cos^{m-1} u}{m+n} + \frac{m-1}{m+n}\int \sin^n u \cos^{m-2} u\, du$

$(n \neq -m)$

18. $\displaystyle\int x^n \sin ax\, dx = -\frac{x^n}{a}\cos ax + \frac{n}{a}\int x^{n-1}\cos ax\, dx$

19. $\displaystyle\int x^n \cos ax\, dx = \frac{x^n}{a}\sin ax - \frac{n}{a}\int x^{n-1}\sin ax\, dx$

20. $\displaystyle\int \tan^n ax\, dx = \frac{1}{a}\frac{\tan^{n-1} ax}{n-1} - \int \tan^{n-2} ax\, dx \quad (n \neq 1)$

Formulas Involving Quadratic Expressions

21. $\displaystyle \int \frac{du}{a^2 + u^2} = \frac{1}{a}\tan^{-1}\frac{u}{a}$

22. $\displaystyle \int \frac{du}{(a^2 + u^2)^2} = \frac{u}{2a^2(a^2 + u^2)} + \frac{1}{2a^3}\tan^{-1}\frac{u}{a}$

23. $\displaystyle \int \frac{du}{a^2 - u^2} = \frac{1}{2a}\ln\left|\frac{u + a}{u - a}\right|$

24. $\displaystyle \int \frac{du}{(a^2 - u^2)^2} = \frac{u}{2a^2(a^2 - u^2)} + \frac{1}{4a^3}\ln\left|\frac{u + a}{u - a}\right|$

25. $\displaystyle \int \frac{du}{\sqrt{a^2 + u^2}} = \ln(u + \sqrt{a^2 + u^2})$

26. $\displaystyle \int \sqrt{a^2 + u^2}\,du = \frac{u}{2}\sqrt{a^2 + u^2} + \frac{a^2}{2}\ln(u + \sqrt{a^2 + u^2})$

27. $\displaystyle \int \frac{du}{\sqrt{a^2 - u^2}} = \sin^{-1}\frac{u}{a}$

28. $\displaystyle \int \sqrt{a^2 - u^2}\,du = \frac{u}{2}\sqrt{a^2 - u^2} + \frac{a^2}{2}\sin^{-1}\frac{u}{a}$

29. $\displaystyle \int \frac{du}{\sqrt{u^2 - a^2}} = \ln\left|u + \sqrt{u^2 - a^2}\right|$

30. $\displaystyle \int \sqrt{u^2 - a^2}\,du = \frac{u}{a}\sqrt{u^2 - a^2} - \frac{a^2}{2}\ln\left|u + \sqrt{u^2 - a^2}\right|$

Formulas Involving the Exponential Function

31. $\displaystyle\int e^u \, du = e^u$

32. $\displaystyle\int x e^{ax} \, dx = \frac{1}{a^2}(ax - 1)e^{ax}$

33. $\displaystyle\int x^n e^{ax} \, dx = \frac{1}{a}x^n e^{ax} - \frac{n}{a}\int x^{n-1} e^{ax} \, dx$

34. $\displaystyle\int e^{ax} \sin bx \, dx = \frac{e^{ax}}{a^2 + b^2}(a \sin bx - b \cos bx)$

35. $\displaystyle\int e^{ax} \cos bx \, dx = \frac{e^{ax}}{a^2 + b^2}(a \cos bx + b \sin bx)$

36. $\displaystyle\int_0^\infty e^{-au^2} \, du = \frac{1}{2}\sqrt{\frac{\pi}{a}}$

37. $\displaystyle\int_0^\infty u^n e^{-u} \, du = n!$

38. $\displaystyle\int x^n \ln ax \, dx = \frac{x^{n+1}}{(n+1)^2}[(n+1)\ln ax - 1] \quad (n \neq -1)$

39. $\displaystyle\int a^u \, du = \frac{1}{\ln a}a^u$

www.ingramcontent.com/pod-product-compliance
Lightning Source LLC
Chambersburg PA
CBHW051442170526
45166CB00001B/79